Collins

ROYAL
OBSERVATORY
GREENWICH

2024 GUIDE
to the
NIGHT SKY

Storm Dunlop and Wil Tirion

T0173720

Published by Collins
An imprint of HarperCollins Publishers
Westerhill Road, Bishopbriggs
Glasgow G64 2QT

HarperCollins Publishers
Macken House,
39/40 Mayor Street Upper,
Dublin 1, D01 C9W8, Ireland
www.harpercollins.co.uk

In association with
Royal Museums Greenwich, the group name for the National Maritime Museum,
Royal Observatory Greenwich, Queen's House and *Cutty Sark* 2015
www.rmg.co.uk

© HarperCollins Publishers 2023
Text and illustrations © Storm Dunlop and Wil Tirion
Photographs © see acknowledgements page 110

A catalogue record for this book is available from the British Library

ISBN 978-0-00-861962-6

10 9 8 7 6 5 4 3 2 1

Printed in the UK

If you would like to comment on any aspect of this book, please contact us at the above address or online.

Contents

Introduction

The aim of this Guide is to help people find their way around the night sky, by showing how the stars that are visible change from month to month and by including details of various events that occur throughout the year. The objects and events described may be observed with the naked eye, or nothing more complicated than a pair of binoculars.

The conditions for observing naturally vary over the course of the year. During the summer, twilight may persist throughout the night and make it difficult to see the faintest stars. There are three recognized stages of twilight: civil twilight, when the Sun is less than 6° below the horizon; nautical twilight, when the Sun is between 6° and 12° below the horizon; and astronomical twilight, when the Sun is between 12° and 18° below the horizon. Full darkness occurs only when the Sun is more than 18° below the horizon. During nautical twilight, only the very brightest stars are visible. During astronomical twilight, the faintest stars visible to the naked eye may be seen directly overhead, but are lost at lower altitudes. As the diagram shows, during most of June full darkness never occurs at the latitude of Vancouver, BC. Slightly farther south (at Seattle, for example) it is truly dark for about two hours, and for somewhere like Houston, TX, there are at least six hours of darkness.

Another factor that affects the visibility of objects is the amount of moonlight in the sky. At Full Moon, it may be very difficult to see some of the fainter stars and objects, and even when the Moon is at a smaller phase it may seriously interfere with visibility if it is near the stars or planets in which you are interested. A full lunar calendar is given for each month and may be used to see when nights are likely to be darkest and best for observation.

The celestial sphere

All the objects in the sky (including the Sun, Moon and stars) appear to lie at some indeterminate distance on a large sphere, centered on the Earth. This **celestial sphere** has various reference points and features that are related to those of the Earth. If the Earth's rotational axis is extended, for example, it points to the North and South Celestial Poles, which are thus in line with the North and South Poles on Earth. Similarly, the **celestial**

The duration of twilight throughout the year at Vancouver and Houston.

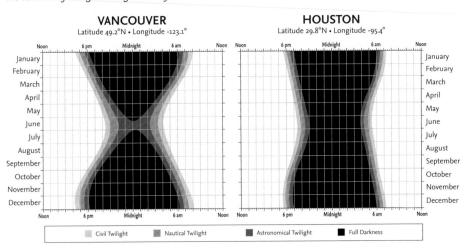

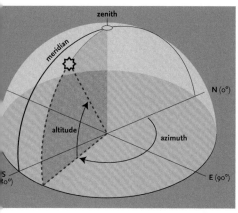

Measuring altitude and azimuth on the celestial sphere.

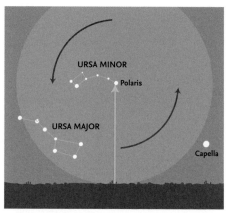

The altitude of the North Celestial Pole equals the observer's latitude.

equator lies in the same plane as the Earth's equator, and divides the sky into northern and southern hemispheres. Because this Guide is written for use in North America, the area of the sky that it describes includes the whole of the northern celestial hemisphere and those portions of the southern that become visible at different times of the year. Stars in the far south, however, remain invisible throughout the year, and are not included.

It is useful to know some of the special terms for various parts of the sky. As seen by an observer, half of the celestial sphere is invisible, below the horizon. The point directly overhead is known as the **zenith**, and the (invisible) one below one's feet as the **nadir**. The line running from the north point on the horizon, up through the zenith and then down to the south point is the **meridian**. This is an important invisible line in the sky because objects are highest in the sky, and thus easiest to see, when they cross the meridian in the south. Objects are said to **transit** when they cross this line in the sky.

In this book, reference is frequently made in the text and in the diagrams to the standard compass points around the horizon. The position of any object in the sky may be described by its **altitude** (measured in degrees above the horizon), and its **azimuth** (measured in degrees from north 0°, through east 90°, south 180° and west 270°). Experienced amateurs and professional astronomers also use another system of specifying locations on the celestial sphere, but that need not concern us here, where the simpler method will suffice.

The celestial sphere appears to rotate about an invisible axis, running between the North and South Celestial Poles. The location (i.e., the altitude) of the Celestial Poles depends entirely on the observer's position on Earth or, more specifically, their latitude. The charts in this book are produced for the latitude of 40°N, so the North Celestial Pole (NCP) is 40° above the northern horizon. The fact that the NCP is fixed relative to the horizon means that all the stars within 40° of the pole are always above the horizon and may, therefore, always be seen at night, regardless of the time of year. This northern circumpolar region is an ideal place to begin learning the sky, and ways to identify the circumpolar stars and constellations will be described shortly.

The ecliptic and the zodiac

Another important line on the celestial sphere is the Sun's apparent path against the background stars – in reality the result of the Earth's orbit around the Sun. This is known as the **ecliptic**. The point where the Sun,

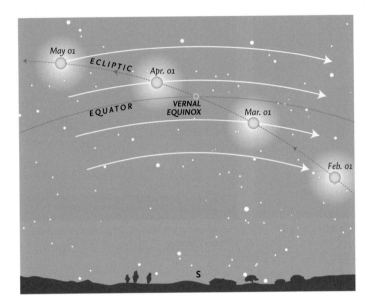

The Sun crossing the celestial equator in spring.

apparently moving along the ecliptic, crosses the celestial equator from south to north is known as the vernal (or spring) equinox, which occurs on March 20. At this time (and at the autumnal equinox, on September 22 or 23, when the Sun crosses the celestial equator from north to south) day and night are almost exactly equal in length. (There is a slight difference, but that need not concern us here.) The vernal equinox is currently located in the constellation of Pisces, and is important in astronomy because it defines the zero point for a system of celestial coordinates, which is, however, not used in this Guide.

The Moon and planets are to be found in a band of sky that extends 8° on either side of the ecliptic. This is because the orbits of the Moon and planets are inclined at various angles to the ecliptic (i.e., to the plane of the Earth's orbit). This band of sky is known as the zodiac and, when originally devised, consisted of 12 **constellations**, all of which were considered to be exactly 30° wide. When the constellation boundaries were formally established by the International Astronomical Union in 1930, the exact extent of most constellations was altered and, nowadays,

the ecliptic passes through 13 constellations. Because of the boundary changes, the Moon and planets may actually pass through several other constellations that are adjacent to the original 12.

The constellations

Since ancient times, the celestial sphere has been divided into various constellations, most dating back to antiquity and usually associated with certain myths or legendary people and animals. Nowadays, the boundaries of the constellations have been fixed by international agreement and their names (in Latin) are largely derived from Greek or Roman originals. Some of the names of the most prominent stars are of Greek or Roman origin, but many are derived from Arabic names. Many bright stars have no individual names and, for many years, stars were identified by terms such as "the star in Hercules' right foot." A more sensible scheme was introduced by the German astronomer Johannes Bayer in the early 17th century. Following his scheme — which is still used today — most of the brightest stars are identified by a Greek letter followed by the genitive form of the constellation's Latin name. An example is the Pole Star, also known as Polaris and α Ursae Minoris (abbreviated α UMi). The Greek alphabet is shown on page 111 with a list of all the constellations that may be seen from latitude 40°N, together with abbreviations, their genitive forms and English names on page 110. Other naming schemes exist for fainter stars, but are not used in this book.

Asterisms

Apart from the constellations (88 of which cover the whole sky), certain groups of stars, which may form a part of a larger constellation or cross several constellations, are readily recognizable and have been given individual names. These groups are known as *asterisms*, and the most famous (and well-known) is the "Big Dipper," the common name for the seven brightest stars in the constellation of Ursa Major, the Great Bear. The names and details of some asterisms mentioned in this book are given in the list on page 111.

Magnitudes

The brightness of a star, planet or other body is frequently given in magnitudes (mag.). This is a mathematically defined scale where larger numbers indicate a fainter object. The scale extends beyond the zero point to negative numbers for very bright objects. (Sirius, the brightest star in the sky is mag. -1.4.) Most observers are able to see stars to about mag. 6, under very clear skies.

The Moon

Although the daily rotation of the Earth carries the sky from east to west, the Moon gradually moves eastwards by approximately its diameter (about half a degree) in an hour. Normally, in its orbit around the Earth, the Moon passes above or below the direct line between Earth and Sun (at New Moon) or outside the area obscured by the Earth's shadow (at Full Moon). Occasionally, however, the three bodies are more or less perfectly aligned to give an *eclipse*: a solar eclipse at New Moon or a lunar eclipse at Full Moon. Depending on the exact circumstances, a solar eclipse may be merely partial (when the Moon does not cover the whole of the Sun's disk), annular (when the Moon is too far from Earth in its orbit to appear large enough to hide the whole of the Sun), or total. Total and annular eclipses are visible from very restricted areas of the Earth, but partial eclipses are normally visible over a wider area.

Somewhat similarly, at a lunar eclipse, the Moon may pass through the outer zone of the Earth's shadow, the *penumbra* (in a penumbral eclipse, which is not generally perceptible to the naked eye), so that just part of the Moon is within the darkest part of the Earth's shadow, the *umbra* (in a partial eclipse); or completely within the umbra (in a total eclipse). Unlike solar eclipses, lunar eclipses are visible from large areas of the Earth.

Occasionally, as it moves across the sky, the Moon passes between the Earth and individual planets or distant stars, giving rise to an *occultation*. As with solar eclipses, such occultations are visible from restricted areas of the world.

The planets

Because the planets are always moving against the background stars, they are treated in some detail in the monthly pages and information is given when they are close to other planets, the Moon or any of five bright stars that lie near the ecliptic. Such events are known as *appulses* or, more frequently, as *conjunctions*. (There are technical differences in the way these terms are defined and should be used in astronomy, but these need not concern us here.) The positions of the planets are shown for every month on a special chart of the ecliptic.

The term conjunction is also used when a planet is either directly behind or in front of the Sun, as seen from Earth. (Under normal circumstances it will then be invisible.) The conditions of most favorable visibility depend on whether the planet is one of the two known as *inferior planets* (Mercury and Venus) or one of the three *superior planets* (Mars, Jupiter

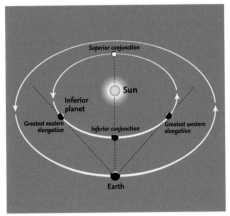

Inferior planet.

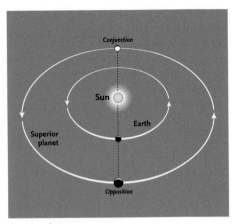

Superior planet.

and Saturn) that are covered in detail. (Some details of the fainter superior planets, Uranus and Neptune, are included in this Guide, and special charts for both are given on page 25.)

The inferior planets are most readily seen at eastern or western **elongation**, when their angular distance from the Sun is greatest. For superior planets, they are best seen at **opposition**, when they are directly opposite the Sun in the sky, and cross the meridian at local midnight.

It is often useful to be able to estimate angles on the sky, and approximate values may be obtained by holding one hand at arm's length. The various angles are shown in the diagram, together with the separations of the various stars in the Big Dipper.

Meteors

At some time or other, nearly everyone has seen a **meteor** – a "shooting star" – as it flashed across the sky. The particles that cause meteors – known technically as "meteoroids" – range in size from that of a grain of sand (or even smaller) to the size of a pea. On any night of the year there are occasional meteors, known as **sporadics**, that may travel in any direction. These occur at a rate that is normally between three and eight in an hour. Far more important, however, are **meteor showers**, which occur at fixed periods of the year, when the

Earth encounters a trail of particles left behind by a comet or, very occasionally, by a minor planet (asteroid). Meteors always appear to diverge from a single point on the sky, known as the **radiant**, and the radiants of major showers are shown on the charts. Meteors that come from a circular area 8° in diameter around the radiant are classed as belonging to the particular shower. All others that do not come from that area are sporadics (or, occasionally from another shower that is active at the same time). A list of the major meteor showers is given on page 31.

Although the positions of the various shower radiants are shown on the charts, looking directly at the radiant is not the most effective way of seeing meteors. They are most likely to be noticed if one is looking about 40°–45° away from the radiant position. (This is approximately two hand-spans as shown in the diagram for measuring angles.)

Other objects

Certain other objects may be seen with the naked eye under good conditions. Some were given names in antiquity – Praesepe is one example – but many are known by what are called "Messier numbers," the numbers in a catalogue of nebulous objects compiled by Charles Messier in the late 18th century. Some, such as the Andromeda Galaxy, M31, and the Orion Nebula, M42, may be seen

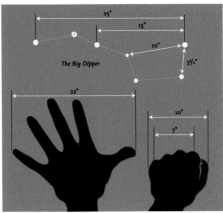

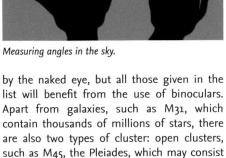

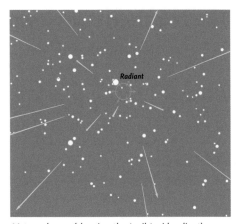

Measuring angles in the sky.

Meteor shower (showing the April Lyrid radiant).

by the naked eye, but all those given in the list will benefit from the use of binoculars. Apart from galaxies, such as M31, which contain thousands of millions of stars, there are also two types of cluster: open clusters, such as M45, the Pleiades, which may consist of a few dozen to some hundreds of stars; and globular clusters, such as M13 in Hercules, which are spherical concentrations of many thousands of stars. One or two gaseous nebulae, consisting of gas illuminated by stars within them, are also visible. The Orion Nebula, M42, is one, and is illuminated by the group of four stars, known as the Trapezium, which may be seen within it by using a good pair of binoculars.

Some interesting objects.

Messier / NGC	Name	Type	Constellation	Maps (months)
—	Hyades	open cluster	Taurus	Sep. – Mar.
—	Double Cluster	open cluster	Perseus	All year
—	Melotte 111 (Coma Cluster)	open cluster	Coma Berenices	Jan. – Aug.
M3	—	globular cluster	Canes Venatici	Feb. – Aug.
M4	—	globular cluster	Scorpius	May – Aug.
M8	Lagoon Nebula	gaseous nebula	Sagittarius	Jun. – Sep.
M11	Wild Duck Cluster	open cluster	Scutum	May – Oct.
M13	Hercules Cluster	globular cluster	Hercules	Mar. – Oct.
M15	—	globular cluster	Pegasus	Jun. – Dec.
M20	Trifid Nebula	gaseous nebula	Sagittarius	Jun. – Sep.
M22	—	globular cluster	Sagittarius	Jun. – Sep.
M27	Dumbbell Nebula	planetary nebula	Vulpecula	May – Nov.
M31	Andromeda Galaxy	galaxy	Andromeda	Jun. – Mar.
M35	—	open cluster	Gemini	Oct. – Apr.
M42	Orion Nebula	gaseous nebula	Orion	Nov. – Mar.
M44	Praesepe	open cluster	Cancer	Nov. – Jun.
M45	Pleiades	open cluster	Taurus	Sep. – Mar.
M57	Ring Nebula	planetary nebula	Lyra	Apr. – Nov.
M67	—	open cluster	Cancer	Dec. – May
NGC 752	—	open cluster	Andromeda	Jul. – Mar.
NGC 3242	Ghost of Jupiter	planetary nebula	Hydra	Feb. – May

The Northern Circumpolar Constellations

The northern circumpolar stars are the key to starting to identify the constellations. For anyone in the northern hemisphere they are visible at any time of the year, and nearly everyone is familiar with the seven stars of the Big Dipper: an asterism that forms part of the large constellation of **Ursa Major** (the Great Bear).

Ursa Major

Because of the movement of the stars caused by the passage of the seasons, Ursa Major lies in different parts of the evening sky at different periods of the year. The diagram below shows its position at the beginning of the four main seasons. The seven stars of the Big Dipper remain visible throughout the year anywhere north of latitude 40°N. Even at the latitude (40°N) for which the charts in this book are drawn, many of the stars in the southern portion of the constellation of Ursa Major are hidden below the horizon for part of the year or (particularly in late summer) cannot be seen late in the night.

Polaris and Ursa Minor

The two stars **Dubhe** and **Merak** (α and β Ursae Majoris, respectively), farthest from the "tail"

are known as the "Pointers." A line from Merak to Dubhe, extended about five times their separation, leads to the Pole Star, **Polaris**, or α Ursae Minoris. All the stars in the northern sky appear to rotate around it. There are five main stars in the constellation of **Ursa Minor**, and the two farthest from the Pole, **Kochab** and **Pherkad** (β and γ Ursae Minoris, respectively), are known as "The Guards."

Cassiopeia

On the opposite of the North Pole from Ursa Major lies **Cassiopeia**. It is highly distinctive, appearing as five stars forming a letter "W" or "M" depending on its orientation. Provided the sky is reasonably clear of clouds, you will nearly always be able to see either Ursa Major or Cassiopeia, and thus be able to orientate yourself on the sky.

To find Cassiopeia, start with **Alioth** (ε Ursae Majoris), the first star in the tail of the Great Bear. A line from this star extended through Polaris points directly toward γ Cassiopeiae, the central star of the five.

Cepheus

Although the constellation of **Cepheus** is fully circumpolar, it is not nearly as well-known as Ursa Major, Ursa Minor or Cassiopeia, partly because its stars are fainter. Its shape is rather like the end of a house with a pointed roof. The line from the Pointers through Polaris, if extended, leads to **Errai** (γ Cephei) at the "top" of the "roof." The brightest star, **Alderamin** (α Cephei) lies in the Milky Way region, at the "bottom right-hand corner" of the figure.

Draco

The constellation of **Draco** consists of a quadrilateral of stars, known as the "Head of Draco" (and also the "Lozenge"), and a long chain of stars forming the neck and body of the dragon. To find the Head of Draco, locate the two stars **Phecda** and **Megrez** (γ and δ Ursae Majoris) in the Big Dipper,

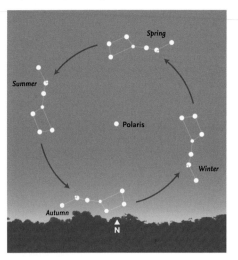

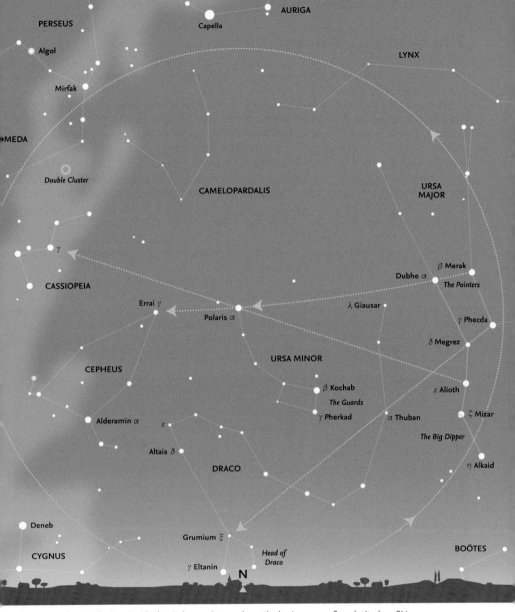

The stars and constellations inside the circle are always above the horizon, seen from latitude 40°N.

opposite the Pointers. Extend a line from Phecda through Megrez by about eight times their separation, right across the sky below the Guards in Ursa Minor, ending at **Grumium** (ξ Draconis) at one corner of the quadrilateral. The brightest star, **Eltanin** (γ Draconis) lies farther to the south. From the head of Draco, the constellation first runs northwest to **Altais** (δ Draconis) and ε Draconis, then doubles back southwards before winding its way through **Thuban** (α Draconis) before ending at **Giausar** (λ Draconis) between the Pointers and Polaris.

The Winter Constellations

The winter sky is dominated by several bright stars and distinctive constellations. The most conspicuous constellation is **Orion**, the main body of which has an hour-glass shape. It straddles the celestial equator and is thus visible from anywhere in the world. The three stars that form the "Belt" of Orion point down towards the southeast and to **Sirius** (α Canis Majoris), the brightest star in the sky. **Mintaka** (δ Orionis), the star at the northeastern end of the Belt, farthest from Sirius, actually lies just slightly south of the celestial equator.

A line from **Bellatrix** (γ Orionis) at the "top right-hand corner" of Orion, through **Aldebaran** (α Tauri), past the "V" of the Hyades cluster, points to the distinctive cluster of bright blue stars known as the **Pleiades**, or the "Seven Sisters." Aldebaran is one of the five bright stars that may sometimes be occulted (hidden) by the Moon. Another line from Bellatrix, through **Betelgeuse** (α Orionis), if carried right across the sky, points to the constellation of **Leo**, a prominent constellation in the spring sky.

Six bright stars in six different constellations: **Capella** (α Aurigae), **Aldebaran** (α Tauri), **Rigel** (β Orionis), **Sirius** (α Canis Majoris), **Procyon** (α Canis Minoris) and **Pollux** (β Gemini) form what is sometimes known as the "Winter Hexagon." Pollux is accompanied to the northwest by the slightly fainter star of **Castor** (α Gemini), the second "Twin."

In a counterpart to the famous "Summer Triangle," an almost perfect equilateral triangle, the "Winter Triangle," is formed by Betelgeuse (α Orionis), Sirius (α Canis Majoris) and Procyon (α Canis Minoris).

Several of the stars in this region of the sky show distinctive tints: Betelgeuse (α Orionis) is reddish, Aldebaran (α Tauri) is orange and Rigel (β Orionis) is blue-white.

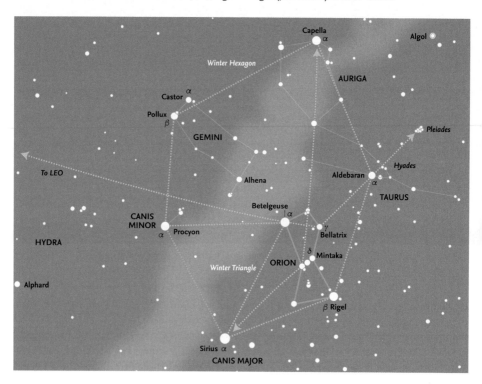

The Spring Constellations

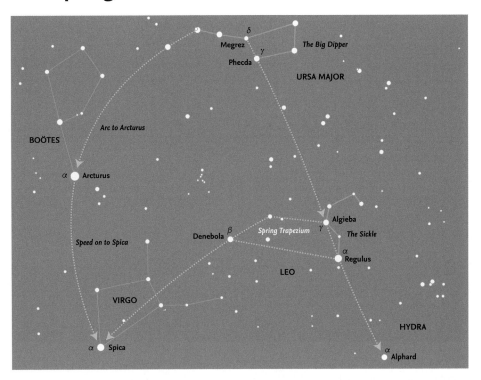

The most prominent constellation in the spring sky is the zodiacal constellation of **Leo**, and its brightest star, **Regulus** (α Leonis), which may be found by extending a line from Megrez and Phecda (δ and γ Ursae Majoris, respectively) – the two stars on the opposite side of the bowl of the Big Dipper from the Pointers – down to the southeast. Regulus forms the "dot" of the "backward question mark" known as "the Sickle." Regulus, like Aldebaran in Taurus is one of the bright stars that lie close to the ecliptic, and that are occasionally occulted by the Moon. The same line from Ursa Major to Regulus, if continued, leads to **Alphard** (α Hydrae), the brightest star in **Hydra**, the largest of the 88 constellations.

The shape formed by the body of Leo is sometimes known as the "Spring Trapezium." At the other end of the constellation from Regulus is **Denebola** (β Leonis), and the line

forming the back of the constellation through Denebola points to the bright star **Spica** (α Virginis) in the constellation of **Virgo**. A saying that helps to locate Spica is well-known to astronomers: "Arc to Arcturus and then speed on to Spica." This suggests following the arc of the tail of Ursa Major to Arcturus and then on to Spica. **Arcturus** (α Boötis) is actually the brightest star in the northern hemisphere of the sky. (Although other stars, such as Sirius, are brighter, they are all in the southern hemisphere.) Overall, the constellation of **Boötes** is sometimes described as "kite-shaped" or "shaped like the letter P."

Although Spica is the brightest star in the zodiacal constellation of Virgo, the rest of the constellation is not particularly distinct, consisting of a rough quadrilateral of moderately bright stars and some fainter lines of stars extending outwards.

The Summer Constellations

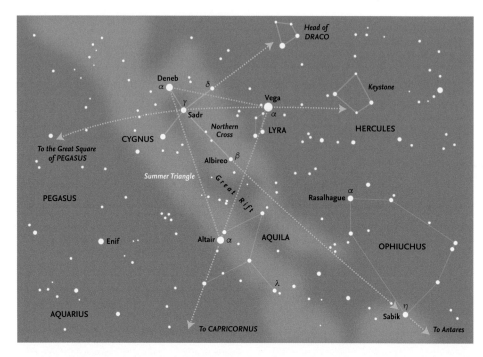

On summer nights, the three bright stars **Deneb** (α Cygni), **Vega** (α Lyrae) and **Altair** (α Aquilae) form the striking "Summer Triangle." The constellations of **Cygnus** (the Swan) and **Aquila** (the Eagle) represent birds "flying" down the length of the Milky Way. This part of the Milky Way contains the **Great Rift**, an elongated dark region, where the light from distant stars is obscured by intervening dust. The dark Rift is clearly visible even to the naked eye.

The most prominent stars of Cygnus are sometimes known as the "Northern Cross" (as a counterpart to the "Southern Cross" – the constellation of Crux – in the southern hemisphere). The central line of Cygnus through **Albireo** (β Cygni), extended well to the southwest, points to **Sabik** (η Ophiuchi) in the large, sprawling constellation of **Ophiuchus** (the Serpent Bearer) and beyond to **Antares** (α Scorpii) in the constellation of **Scorpius**. Like Cepheus, the shape of Ophiuchus somewhat resembles the gable-end of a house, and the brightest star **Rasalhague** (α Ophiuchi) is at the "apex" of the "gable."

A line from the central star of Cygnus, **Sadr** (γ Cygni) through δ Cygni, in the northwestern "wing" points toward the Head of Draco, and is another way of locating that part of the constellation. Another line from Sadr to Vega indicates the central portion, "The Keystone," of the constellation of **Hercules**. An arc through the same stars, in the opposite direction, points toward the constellation of **Pegasus**, and more specifically to the "Great Square of Pegasus."

Aquila is less conspicuous than Cygnus and consists of a diamond shape of stars, representing the body and wings of the eagle, together with a rather faint star, λ Aquilae, marking the "head." **Lyra** (the Lyre) mainly consists of Vega (α Lyrae) and a small quadrilateral of stars to its southeast. Continuation of a line from Vega through Altair indicates the zodiacal constellation of **Capricornus**.

The Autumn Constellations

During the autumn season, the most striking feature is the "Great Square of Pegasus," an almost perfect rectangle on the sky, forming the main body of the constellation of **Pegasus**. However, the star at the northeastern corner, **Alpheratz**, is actually α Andromedae, and part of the adjacent constellation of **Andromeda**. A line from **Scheat** (β Pegasi) at the northeastern corner of the Square, through **Matar** (η Pegasi), points in the general direction of Cygnus. A crooked line of stars leads from **Markab** (α Pegasi) through **Homam** (ζ Pegasi) and **Biham** (θ Pegasi) to **Enif** (ε Pegasi). A line from Markab through the last star in the Square, **Algenib** (γ Pegasi) points in the general direction of the five stars, including **Menkar** (α Ceti) that form the "tail" of the constellation of **Cetus** (the Whale). A ring of seven stars lying below the southern side of the Great Square is known as "the Circlet," part of the constellation of **Pisces** (the Fishes).

Extending the line of the western side of the Great Square toward the south leads to the isolated bright star, **Fomalhaut** (α Piscis Austrini), in the Southern Fish. Following the line of the eastern side of the Great Square toward the north leads to Cassiopeia while, in the other direction, it points toward **Diphda** (β Ceti), which is actually the brightest star in Cetus.

Three bright stars leading northeast from Alpheratz form the main body of the constellation of **Andromeda**. Continuation of that line leads toward the constellation of **Perseus** and **Mirfak** (α Persei). Running southwards from Mirfak is a chain of stars, one of which is the famous variable star **Algol** (β Persei). Farther east, an arc of stars leads to the prominent cluster of the **Pleiades**, in the constellation of **Taurus**.

Between Andromeda and Cetus lie the two small constellations of **Triangulum** (the Triangle) and **Aries** (the Ram).

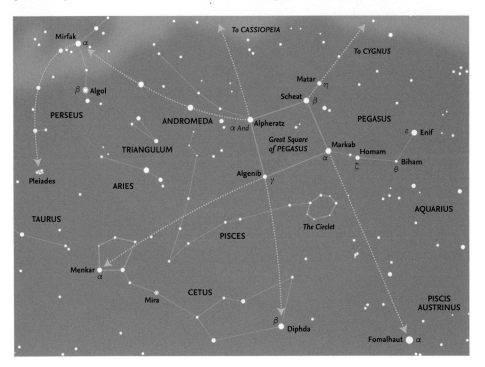

The Moon at First Quarter.

The Moon

The monthly pages include diagrams showing the phase of the Moon for every day of the month, and also indicate the day in the **lunation** (or *age* of the Moon), which begins at New Moon. Although the main features of the surface – the light highlands and the dark maria (seas) – may be seen with the naked eye, far more features may be detected with the use of binoculars or any telescope. The many craters are best seen when they are close to the **terminator** (the boundary between the illuminated and the non-illuminated areas of the surface), when the Sun rises or sets over any particular region of the Moon and the crater walls or central peaks cast strong shadows. Most features become difficult to see at Full Moon, although this is the best time to see the bright ray systems surrounding certain craters. Accompanying the Moon map on the following pages is a list of prominent features, including the days in the lunation when features are normally close to the terminator and thus easiest to see. A few bright features

such as Linné and Proclus, visible when well illuminated, are also listed. One feature, Rupes Recta (the Straight Wall) is readily visible only when it casts a shadow with light from the east, appearing as a light line when illuminated from the opposite direction.

The dates of visibility vary slightly through the effects of **libration**. Because the Moon's orbit is inclined to the Earth's equator and also because it moves in an ellipse, the Moon appears to rock slightly from side to side (and nod up and down). Features near the **limb** (the edge of the Moon) may vary considerably in their location and visibility. (This is easily noticeable with Mare Crisium and the craters Tycho and Plato.) Another effect is that at crescent phases before and after New Moon, the normally non-illuminated portion of the Moon receives a certain amount of light, reflected from the Earth. This **Earthshine** may enable certain bright features (such as Aristarchus, Kepler and Copernicus) to be detected even though they are not illuminated by sunlight.

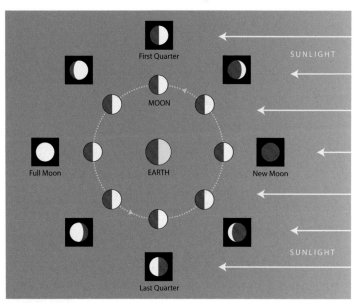

The Moon phases. *During its orbit around the Earth we see different portions of the illuminated side of the Moon's surface.*

Map of the Moon

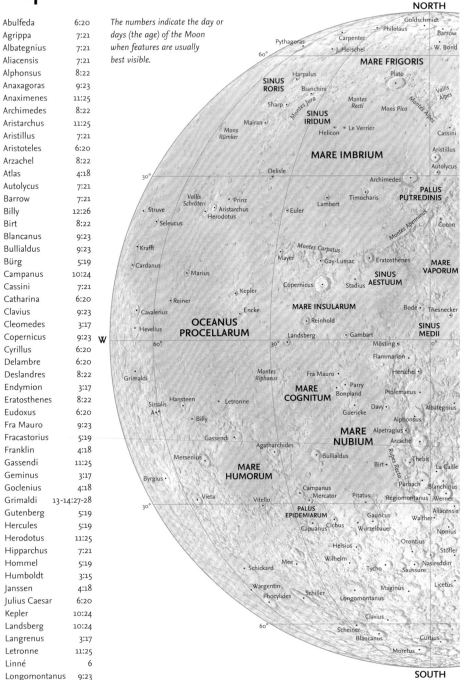

Abulfeda 6:20
Agrippa 7:21
Albategnius 7:21
Aliacensis 7:21
Alphonsus 8:22
Anaxagoras 9:23
Anaximenes 11:25
Archimedes 8:22
Aristarchus 11:25
Aristillus 7:21
Aristoteles 6:20
Arzachel 8:22
Atlas 4:18
Autolycus 7:21
Barrow 7:21
Billy 12:26
Birt 8:22
Blancanus 9:23
Bullialdus 9:23
Bürg 5:19
Campanus 10:24
Cassini 7:21
Catharina 6:20
Clavius 9:23
Cleomedes 3:17
Copernicus 9:23
Cyrillus 6:20
Delambre 6:20
Deslandres 8:22
Endymion 3:17
Eratosthenes 8:22
Eudoxus 6:20
Fra Mauro 9:23
Fracastorius 5:19
Franklin 4:18
Gassendi 11:25
Geminus 3:17
Goclenius 4:18
Grimaldi 13-14:27-28
Gutenberg 5:19
Hercules 5:19
Herodotus 11:25
Hipparchus 7:21
Hommel 5:19
Humboldt 3:15
Janssen 4:18
Julius Caesar 6:20
Kepler 10:24
Landsberg 10:24
Langrenus 3:17
Letronne 11:25
Linné 6
Longomontanus 9:23

The numbers indicate the day or days (the age) of the Moon when features are usually best visible.

NORTH

Goldschmidt
Philolaus
Pythagoras
Carpenter
Barrow
W. Bond
J. Herschel
60°
MARE FRIGORIS
Harpalus
Plato
SINUS RORIS
Bianchini
Vallis Alpes
Sharp
Montes Recti
Montes Alpes
Mons Pico
SINUS IRIDUM
Mairan
Le Verrier
Mons Rümker
Helicon
Cassini
Aristillus
MARE IMBRIUM
Autolycus
Delisle
30°
Archimedes
PALUS PUTREDINIS
Vallis Schröteri
Prinz
Timocharis
Struve
Aristarchus
Lambert
Herodotus
Euler
Seleucus
Montes Apenninus
Cocon
Krafft
Montes Carpatus
MARE VAPORUM
Cardanus
Mayer
Gay-Lussac
Eratosthenes
Marius
SINUS AESTUUM
Copernicus
Stadius
Kepler
Reiner
MARE INSULARUM
Bode
Theisnecker
Cavalerius
Encke
Reinhold
SINUS MEDII
Hevelius
OCEANUS PROCELLARUM
Landsberg
Gambart
W
60°
30°
Mösting
0°
Flammarion
Delambre
Montes Riphaeus
Fra Mauro
Herschel
Grimaldi
MARE COGNITUM
Parry
Ptolemaeus
Sirsalis
Hansteen
Letronne
Bonpland
A
Davy
Albategnius
Billy
Guericke
Alphonsus
MARE NUBIUM
Alpetragius
Gassendi
Arzachel
Agatharchides
Merserius
Bullialdus
Thebit
Birt
La Caille
Rupes Recta
MARE HUMORUM
Campanus
Purbach
Blanchinus
Byrgius
Mercator
Pitatus
Regiomontanus
Werner
Vieta
Vitello
PALUS EPIDEMIARUM
Gauricus
Walther
Aliacensis
Capuanus
Cichus
Wurzelbauer
Nonius
Helsius
Orontius
Wilhelm
Stöfler
Mee
Schickard
Tycho
Nasireddin
Saussure
Wargentin
Maginus
Licetus
Phocylides
Schiller
Longomontanus
Clavius
60°
Scheiner
Blancanus
Curtius
Moretus

SOUTH

18

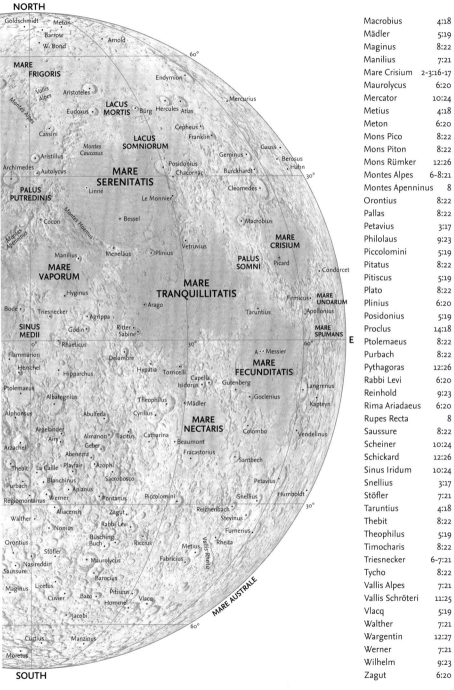

Macrobius 4:18
Mädler 5:19
Maginus 8:22
Manilius 7:21
Mare Crisium 2-3:16-17
Maurolycus 6:20
Mercator 10:24
Metius 4:18
Meton 6:20
Mons Pico 8:22
Mons Piton 8:22
Mons Rümker 12:26
Montes Alpes 6-8:21
Montes Apenninus 8
Orontius 8:22
Pallas 8:22
Petavius 3:17
Philolaus 9:23
Piccolomini 5:19
Pitatus 8:22
Pitiscus 5:19
Plato 8:22
Plinius 6:20
Posidonius 5:19
Proclus 14:18
Ptolemaeus 8:22
Purbach 8:22
Pythagoras 12:26
Rabbi Levi 6:20
Reinhold 9:23
Rima Ariadaeus 6:20
Rupes Recta 8
Saussure 8:22
Scheiner 10:24
Schickard 12:26
Sinus Iridum 10:24
Snellius 3:17
Stöfler 7:21
Taruntius 4:18
Thebit 8:22
Theophilus 5:19
Timocharis 8:22
Triesnecker 6-7:21
Tycho 8:22
Vallis Alpes 7:21
Vallis Schröteri 11:25
Vlacq 5:19
Walther 7:21
Wargentin 12:27
Werner 7:21
Wilhelm 9:23
Zagut 6:20

Eclipses in 2024

Lunar eclipses

There are two lunar eclipses in 2024, one penumbral on March 25, with one partial eclipse on September 18. The penumbral eclipse is of minimal interest, mainly because to the naked eye it will show little variation in the Moon. During the first eclipse, on March 25, a very small portion of the Moon may just graze the umbra. Maximum eclipse is at 07:12. The event (such as it is) will be visible in its totality from North America. During the second (partial) eclipse, on September 18, a slightly larger portion of the Moon will enter the umbra, beginning at 02:14 and leaving at 03:16, with maximum eclipse at 02:44. The middle and end of this eclipse will be just visible from the eastern seaboard of the United States.

Solar eclipses

There are two solar eclipses in 2024. The total eclipse of April 8 is the most significant. The path of totality passes across the southern United States and is shown on the map below. Maximum eclipse actually occurs in Mexico at 18:17, where the duration is 4 minutes 28.1 seconds, but the path of totality runs all the way from Texas to Nova Scotia and Newfoundland in Canada, ending in the North Atlantic.

The second solar eclipse on October 2 is an annular eclipse, mainly visible over the Pacific Ocean, although the end of the track does pass across southern Chile and Argentina. Although most of the track passes over the open ocean, it will cross Rapa Nui (Easter Island, at 27°7'S, 109°22'W) where the annular phase will be visible and last over six minutes. Maximum eclipse occurs at 18:46.

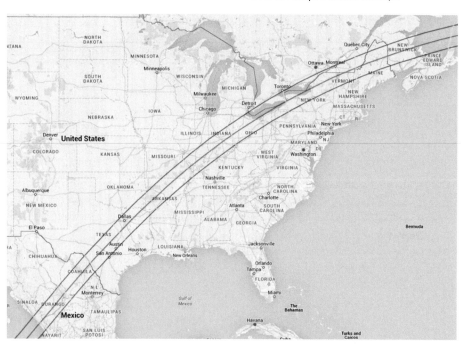

The total solar eclipse of April 8, 2024. The path of totality sees maximum eclipse duration occur in Mexico, before the track enters the United States in Texas. The path runs right across the United States, with the ground track ending in Nova Scotia and Newfoundland in Canada.

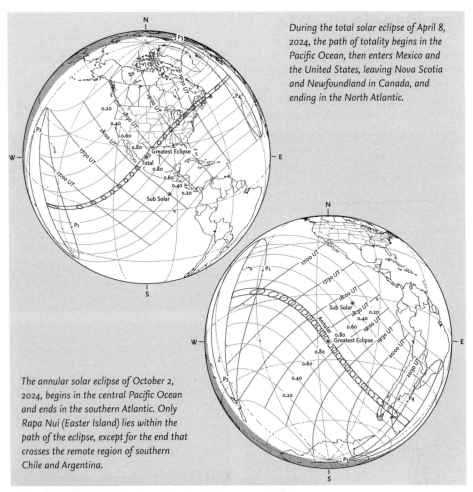

During the total solar eclipse of April 8, 2024, the path of totality begins in the Pacific Ocean, then enters Mexico and the United States, leaving Nova Scotia and Newfoundland in Canada, and ending in the North Atlantic.

The annular solar eclipse of October 2, 2024, begins in the central Pacific Ocean and ends in the southern Atlantic. Only Rapa Nui (Easter Island) lies within the path of the eclipse, except for the end that crosses the remote region of southern Chile and Argentina.

An annular eclipse of the Sun, photographed from California, towards the end of the eclipse at sunset over the Pacific.

The Planets in 2024

Mercury and Venus

Although **Mercury** comes to greatest elongation on seven occasions in 2024, on only six is it reasonably visible above the horizon: January 12, March 24, July 22 September 5, November 16 and December 25. On three of these (January 12, July 22 and November 16), Venus is nearby. **Venus** comes to greatest elongation on October 23, 2023 and January 10, 2025, but not in 2024.

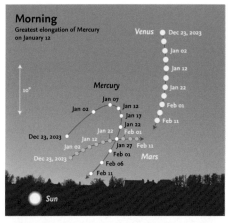

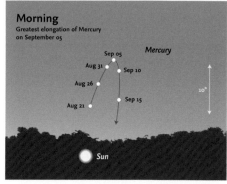

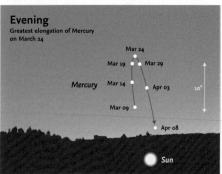

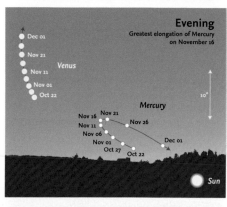

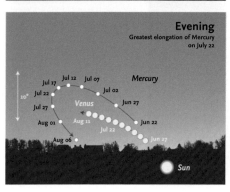

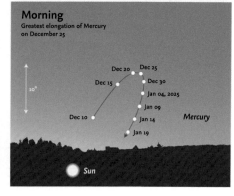

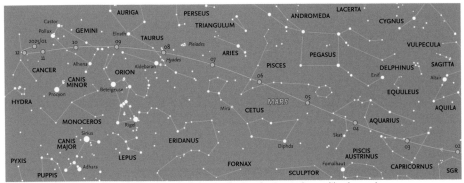

The path of Mars in 2024. In the first months it is too close to the Sun to be readily observed.

Mars

Because its orbit lies outside that of the Earth, so taking longer to complete an orbit, Mars does not come to opposition every year. Instead it often remains in the sky for many months at a time, moving slowly along the ecliptic. This is the case in 2024, when there is no opposition, but the planet is visible from March to December, as shown in the large chart here. (In January and February it is too close to the Sun to be readily visible.) It is mag. 1.3 in *Capricornus* in March, but slowly increases to mag. 1.0 in June and July, in *Pisces* and *Aries*. It ends the year in *Cancer* at mag. -1.2.

Oppositions occur during a period of retrograde motion, when the planet appears to move westwards against the pattern of distant stars. There is no opposition in 2024. Mars actually begins to retrograde on December 9, 2024, and comes to its next opposition on January 16, 2025.

Because of its eccentric orbit, which carries it at very differing distances from the Sun (and Earth), not all oppositions of Mars are equally favorable for observation. The relative positions of Mars and the Earth are shown here. It will be seen that the opposition of 2018 was very close and thus favorable for observation, and that of 2020 was also reasonably good. By comparison, opposition in 2027 will be at a far greater distance, so the planet will appear much smaller.

This image of Mars was obtained by Damian Peach with a 14-inch Celestron telescope on September 30, 2020 when Mars was magnitude -2.5. South is at the top.

The oppositions of Mars between 2018 and 2033. As the illustration shows, there is no opposition in 2024.

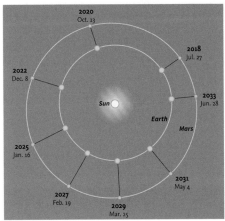

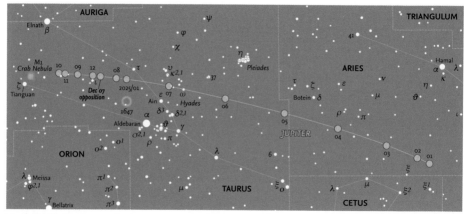

The path of Jupiter in 2024. Jupiter comes to opposition on December 7. Background stars are shown down to magnitude 6.5.

Jupiter and Saturn

In 2024, **Jupiter** starts the year in **Aries**, begining in the west. It moves into **Taurus** in late April. It begins to retrograde on October 9 and comes to opposition, still in Taurus, on December 7 at mag. -2.8. It continues retrograde motion to the end of the year. **Saturn** is in **Aquarius** and stays there for the whole year. It starts retrograde motion on 4 July and comes to opposition at mag. 0.6 on September 8. It continues slow retrograde motion until November 16. At the end of the year it is mag. 1.1.

Jupiter's four large satellites are readily visible in binoculars. Not all four are visible all the time, sometimes hidden behind the planet or invisible in front of it. **Io**, the closest to Jupiter, orbits in just under 1.8 days, and **Callisto**, the farthest away, takes about 16.7 days. In between are **Europa** (c. 3.6 days) and **Ganymede**, the largest (c. 7.1 days). The diagram (below) shows the satellites' motions around the time of opposition.

The path of Saturn in 2024. Saturn comes to opposition on September 8. Background stars are shown down to magnitude 6.5.

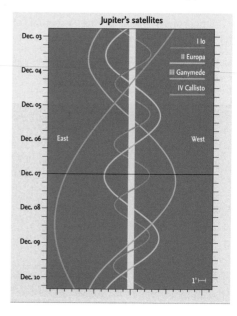

Uranus and Neptune

Uranus begins the year in **Aries** at mag. 5.7, and moves into **Taurus** in late May. It begins retrograde motion in early September and is at opposition (mag. 5.6) on November 17. It continues retrograde motion and is on the **Taurus/Aries** border at the end of the year.

Neptune begins the year just north of the border between **Aquarius** and **Pisces**. It starts its retrograde motion in early July, and is at opposition (mag. 7.8) on September 21. Still in **Pisces** (at mag. 7.9), it reverts to direct motion at the very end of the year.

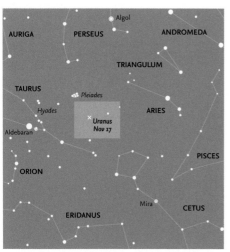

Uranus is in Taurus at the time of its opposition, on November 17. The boxed area is shown in more detail to the right.

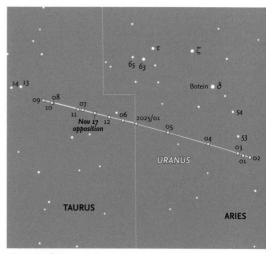

The path of Uranus in 2024. Uranus comes to opposition on November 17. All stars brighter than magnitude 7.5 are shown.

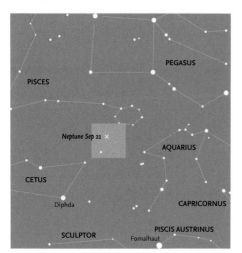

In 2024, Neptune is to be found in the southern part of Pisces. The planet comes to opposition on September 21.

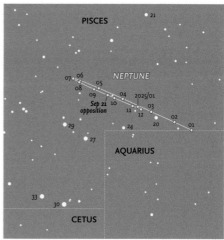

The path of Neptune in 2024. It is in the constellation of Pisces, for the whole year. All stars down to magnitude 8.5 are shown.

Minor Planets in 2024

Several minor planets rise above magnitude 9 in 2024, and four come to opposition at, or above mag. 9.0. These are *(3) Juno* in *Leo* on March 3 when it is mag. 8.7; *(1) Ceres* in *Sagittarius* on July 6, at mag. 7.3, *(7) Iris* in *Aquarius* on August 6 (mag. 8.1) and *(15) Eunomia* in *Auriga* at mag. 8.0 on December 14.

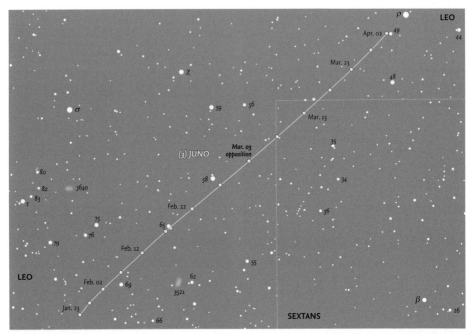

The path of the minor planet (3) Juno around its opposition on March 3 (mag. 8.7). Background stars are shown down to magnitude 9.5.

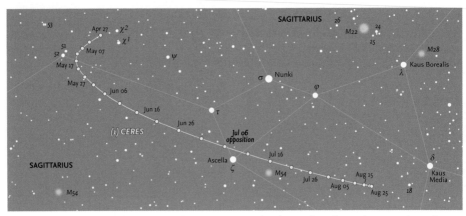

The path of the minor planet (1) Ceres around its opposition on July 6 (mag. 7.3). Background stars are shown down to magnitude 8.5.

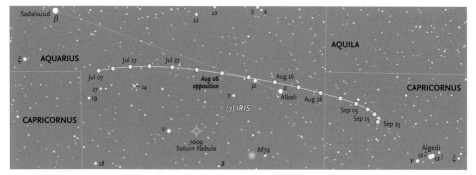

The path of the minor planet (7) Iris around its opposition on August 6 (mag. 8.1). Background stars are shown down to magnitude 9.0.

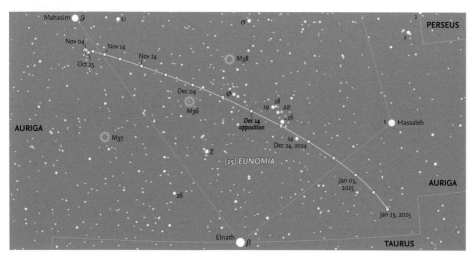

The path of the minor planet (15) Eunomia around its opposition on December 14 (mag. 8.0). Background stars are shown down to magnitude 9.0.

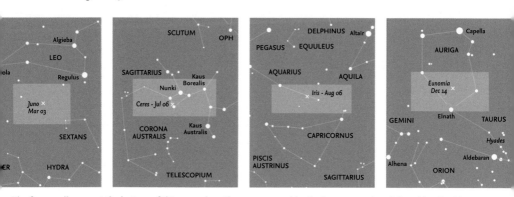

The four small maps at the bottom of this page show the areas covered by the larger maps in a lighter blue. A white cross marks the position of the minor planet at the day of its opposition.

Comets in 2024

Although comets may occasionally become very striking objects in the sky, as was the case with Comet C/2020 F3 NEOWISE in 2020, their occurrence and particularly the existence or length of any tail and their overall magnitude are notoriously difficult to predict. Naturally, it is only possible to predict the return of periodic comets (whose names have the prefix "P"). Many comets appear unexpectedly (these have names with the prefix "C"). Bright, readily visible comets such as C/1995 Y1 Hyakutake, C/1995 O1 Hale-Bopp, C/2006 P1 McNaught or C/2020 F3 NEOWISE are rare. Most periodic comets are faint and only a very small number ever become bright enough to be easily visible with the naked eye or with binoculars.

Comet 62P/Tsuchinshan 1, visible in November and December 2023, may continue to be seen (although fading) in early 2024.

The comet most likely to become readily visible in 2024 is Comet 12P/Pons-Brooks, which may reach mag. 5 during late March or even mag. 4 at the beginning of April. This is a "Halley-type" comet, with an orbital period of 68.82 years. It last came to perihelion on May 22, 1954.

Another comet, C/2021 S3 (PanSTARRS) may rise as high as mag. 7 during April, but is unlikely to become a prominent object.

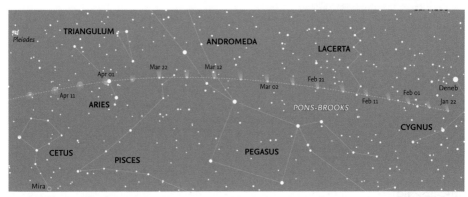

The path of Comet Pons-Brooks from January 22 to April 16, 2024. It rises above magnitude 4 in April. Background stars are shown down to magnitude 6.0.

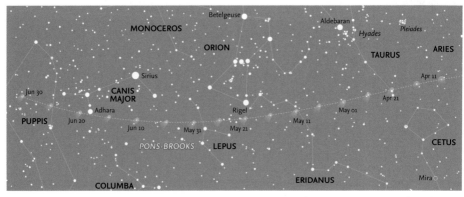

The path of Comet Pons-Brooks from April 11 to June 30, 2024. It rises above magnitude 4 in April. Background stars are shown down to magnitude 6.0.

The Comet C/2020 F3 NEOWISE photographed by Nick James on 17 July 2020, showing the light-colored dust tail, together with the blue ion tail.

Introduction to the Month-by-Month Guide

The monthly charts

The pages devoted to each month contain a pair of charts showing the appearance of the night sky, looking north and looking south. The charts (as with all the charts in this book) are drawn for the latitude of 40°N, so observers farther north will see slightly more of the sky on the northern horizon, and slightly less on the southern. These areas are, of course, those most likely to be affected by poor observing conditions caused by haze, mist or smoke. In addition, stars close to the horizon are always dimmed by atmospheric absorption, so sometimes the faintest stars marked on the charts may not be visible.

The three times shown for each chart require a little explanation. The charts are drawn to show the appearance at 11 p.m. for the 1st of each month. The same appearance will apply an hour earlier (10 p.m.) on the 15th, and yet another hour earlier (9 p.m.) at the end of the month (shown as the 1st of the following month). These times are local time, and apply in all time zones. Where Daylight Saving Time (DST) is used, in 2024 it is introduced on March 10 and ends on November 3. When used, both local time and DST are shown on the charts. Times of specific events are shown in the 24-hour clock of Universal Time (UT) used by astronomers worldwide, and the correct zone time (and DST, where appropriate) may be found from the details inside the front cover.

The charts may be used for earlier or later times during the night. To observe two hours earlier, use the charts for the preceding month; for two hours later, the charts for the next month.

Meteors

Details of specific meteor showers are given in the months when they come to maximum, regardless of whether they begin or end in other months. Note that not all the respective radiants are marked on the charts for that particular month, because the radiants may

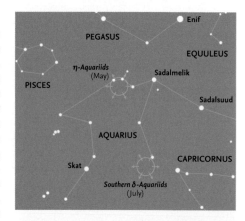

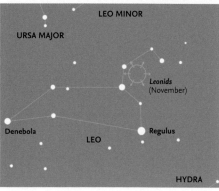

be below the horizon, or lie in constellations that are not readily visible during the month of maximum. For this reason, special charts for the Eta (η) and Delta (δ) Aquariids (May and July, respectively) and the Leonids (November) are shown above. As explained earlier, however, meteors from such showers may still be seen, because the most effective region for seeing meteors is some 40°–45° away from the radiant, and that area of sky may well be above the horizon. A table of the best meteor showers visible during the year is also given here. The rates given are based on the properties of the meteor streams, and are those that an experienced observer might see under ideal conditions. Generally, the observed rates will be far less.

Shower	Dates of activity 2024	Date of maximum 2024	Possible hourly rate
Quadrantids	December 28 to January 12	January 3–4	110
April Lyrids	April 14–30	April 22–23	18
η-Aquariids	April 19 to May 28	May 6	50
α-Capricornids	July 3 to August 15	July 30	5
Southern δ-Aquariids	July 12 to August 23	July 30	25
Perseids	July 17 to August 24	August 12–13	100
α-Aurigids	August 28 to September 5	September 1	6
Southern Taurids	September 10 to November 20	October 10–11	5
Orionids	October 2 to November 7	October 21–22	25
Draconids	October 6–10	October 8–9	10
Northern Taurids	October 20 to December 10	November 12–13	5
Leonids	November 6–30	November 18	10
Geminids	December 4–20	December 14–15	150
Ursids	December 17–26	December 23	10

Meteors that are brighter than magnitude -4 (approximately the maximum magnitude reached by Venus) are known as **fireballs** or **bolides**. Examples are shown on pages 32, 35 and 77. Fireballs sometimes cause sonic booms that may be heard some time after the meteor is seen.

The photographs
As an aid to identification – especially as some people find it difficult to relate charts to the actual stars they see in the sky – one or more photographs of constellations visible in certain specific months are included. It should be noted, however, that because of the limitations of the photographic and printing processes, and the differences between the sensitivity of different individuals to faint starlight (especially in their ability to detect different colors), and the degree to which they have become adapted to the dark, the apparent brightness of stars in the photographs will not necessarily precisely match that seen by any one observer.

The Moon calendar
The Moon calendar is largely self-explanatory. It shows the phase of the Moon for every day of the month, with the exact times (in Universal Time) of New Moon, First Quarter, Full Moon and Last Quarter. Because the times are calculated from the Moon's actual orbital parameters, some of the times shown

will, naturally, fall during daylight, but any difference is too small to affect the appearance of the Moon on that date. Also shown is the **age** of the Moon (the day in the **lunation**), beginning at New Moon, which may be used to determine the best time for observation of specific lunar features.

The Moon
The section on the Moon includes details of any lunar or solar eclipses that may occur during the month (visible from anywhere on Earth). Similar information is given about any important occultations. Mainly, however, this section summarizes when the Moon passes close to planets or the five prominent stars close to the ecliptic. The dates when the Moon is closest to the Earth (at **perigee**) and farthest from it (at **apogee**) are shown in the monthly calendars, and only mentioned here when they are particularly significant, such as the nearest and farthest during the year.

The planets and minor planets
Brief details are given of the location, movement and brightness of the planets from Mercury to Saturn throughout the month. None of the planets can, of course, be seen when they are close to the Sun, so such periods are generally noted. All of the planets may sometimes lie on the opposite side of the Sun to the Earth (at superior conjunction),

A fireball (with flares approximately as bright as the Full Moon), photographed against a weak auroral display by D. Buczynski from Tarbat Ness in Scotland on January 22, 2017.

but in the case of the inferior planets, Mercury and Venus, they may also pass between the Earth and the Sun (at inferior conjunction) and are normally invisible for a period of time, the length of which varies from conjunction to conjunction. Those two planets are normally easiest to see around either eastern or western elongation, in the evening or morning sky, respectively. Not every elongation is favorable, so although every elongation is listed, only those where observing conditions are favorable are shown in the individual diagrams of events.

The dates at which the superior planets reverse their motion (from direct motion to *retrograde*, and retrograde to direct) and of opposition (when a planet generally reaches its maximum brightness) are given. Some planets, especially distant Saturn, may spend most or all of the year in a single constellation. Jupiter and Saturn are normally easiest to see around opposition, which occurs every year. Mars, by contrast, moves relatively rapidly against the background stars and in some years never comes to opposition.

Uranus is normally magnitude 5.7–5.9, and thus at the limit of naked-eye visibility under exceptionally dark skies, but bright enough to be readily visible in binoculars. Because its orbital period is so long (over 84 years), Uranus moves only slowly along the ecliptic, and often remains within a single constellation for a whole year. The chart on page 25 shows its position during 2024.

Similar considerations apply to Neptune, although it is always fainter (generally magnitude 7.8–8.0), still visible in most binoculars. It takes about 164.8 years to complete one orbit of the Sun. As with Uranus, it frequently spends a complete year in one constellation. Its chart is also on page 25.

In any year, few minor planets ever become bright enough to be detectable in binoculars. Just one, (4) Vesta, on rare occasions brightens sufficiently for it to be visible to the naked eye. Our limit for visibility is magnitude 9.0 and details and charts are given for those objects that exceed that magnitude during the year, normally around opposition. To assist in recognition of a planet or minor planet as it moves against the background stars, the latter are shown to a fainter magnitude than the object at opposition. Minor-planet charts for 2024 are on pages 26 and 27.

The ecliptic charts

Although the ecliptic charts are primarily designed to show the positions and motions of the major planets, they also show the motion of the Sun during the month. The light-tinted area shows the area of the sky that is invisible during daylight, but the darker area gives an indication of which constellations are likely to be visible at some time of the night. The closer a planet is to the border between dark and light, the more difficult it will be to see in the twilight.

The monthly calendar

For each month, a calendar shows details of significant events, including when planets are close to one another in the sky, close to the Moon, or close to any one of five bright stars that are spaced along the ecliptic. The times shown are given in Universal Time (UT), always used by astronomers throughout the year. The tables given inside the front cover may be used to convert UT to that used in any given time zone and (during the summer) to DST.

The diagrams of interesting events

Each month, a number of diagrams show the appearance of the sky when certain events take place. However, the exact positions of celestial objects and their separations greatly depend on the observer's position on Earth. When the Moon is one of the objects involved, because it is relatively close to Earth, there may be very significant changes from one location to another. Close approaches between planets or between a planet and a star are less affected by changes of location, which may thus be ignored.

The diagrams showing the appearance of the sky are drawn for the latitude of 40°N and 90°W, northwest of Springfield, IL, so will be approximately correct for most of North America. However, for an observer farther north (say, Vancouver), a planet or star listed as being north of the Moon will appear even farther north, whereas one south of the Moon will appear closer to it – or may even be hidden (occulted) by it. For an observer at a latitude of less than 40°N, there will be corresponding changes in the opposite direction: for a star or planet south of the Moon, the separation will increase, and for one north of the Moon, the separation will decrease. This is particularly important when occultations occur, which may be visible from one location, but not another. In 2024 there are several occultations of Saturn. Only one, on September 17, is visible from North America.

Ideally, details should be calculated for each individual observer, but this is obviously impractical. In fact, positions and separations are actually calculated for a theoretical observer located at the center of the Earth.

So the details given regarding the positions of the various bodies should be used as a guide to their location. A similar situation arises with the times that are shown. These are calculated according to certain technical criteria, which need not concern us here. However, they do not necessarily indicate the exact time when two bodies are closest together. Similarly, dates and times are given, even if they fall in daylight, when the objects are likely to be completely invisible. However, such times do give an indication that the objects concerned will be in the same general area of the sky during both the preceding and the following nights.

Data used in this Guide

The data given in this Guide, such as timings and distances between objects, have been computed with a program developed by the US Naval Observatory in Washington DC (MICA, the Multiyear Interactive Computer Almanac), widely regarded as the most accurate computation. As such, the data may differ slightly from information given elsewhere.

Key to the symbols used on the monthy star maps.

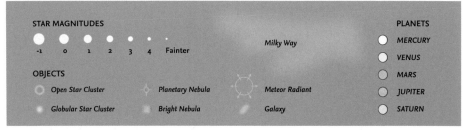

THE MONTH-BY-MONTH GUIDE

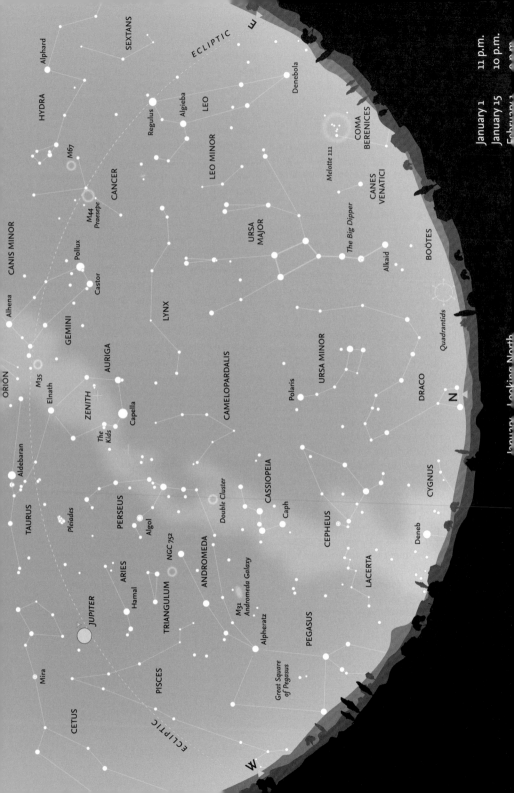

January, Looking North

January 1   11 p.m.
January 15   10 p.m.
February 1   9 p.m.

January – Looking North

Most of the important circumpolar constellations are easy to see in the northern sky at this time of year. *Ursa Major* stands more-or-less vertically above the horizon in the northeast, with the zodiacal constellation of *Leo* rising in the east. To the north, the stars of *Ursa Minor* lie below *Polaris* (the Pole Star). The head of *Draco* is low on the northern horizon, but may be difficult to see unless observing conditions are good. Both *Cepheus* and *Cassiopeia* are readily visible in the northwest, and even the faint constellation of *Camelopardalis* is high enough in the sky for it to be easily visible.

Near the zenith is the constellation of *Auriga* (the Charioteer), with brilliant *Capella* (α Aurigae), directly overhead. Slightly to the west of Capella lies a small triangle of fainter stars, known as "The Kids." (Ancient mythological representations of Auriga show him carrying two young goats.) Together with the northernmost bright star in *Taurus, Elnath* (β Tauri), the body of Auriga forms a large pentagon on the sky, with The Kids lying on the western side. Farther down toward the west are the constellations of *Perseus* and *Andromeda*, and the Great Square of *Pegasus* is approaching the horizon.

Meteors

The *Quadrantid* meteor shower actually begins in late December (on 28 December 2023) and continues until January 12. It is one of the strongest and most consistent meteor showers. It comes to maximum on the night of January 3 to 4, around Last Quarter, so moonlight will cause some interference. However, the meteors are bright, bluish- or yellowish-white and may reach a maximum rate of 110 per hour. The parent object is minor planet 2003 EH_1.

The shower is named after the former constellation *Quadrans Muralis* (the Mural Quadrant), an early form of astronomical instrument. The Quadrantid meteor radiant, marked on the chart, is

A late Quadrantid fireball, photographed from Portmahomack, Ross-shire, Scotland on January 15, 2018 at 23:44 UT.

Ursa Major, the key constellation to navigating the northern sky.

now within the northernmost part of *Boötes*, roughly halfway between θ Boötis and τ Herculis. This is very low on the northern horizon, roughly halfway between *Alkaid*, the last star in the "handle" of the *Big Dipper* and the head of *Draco*.

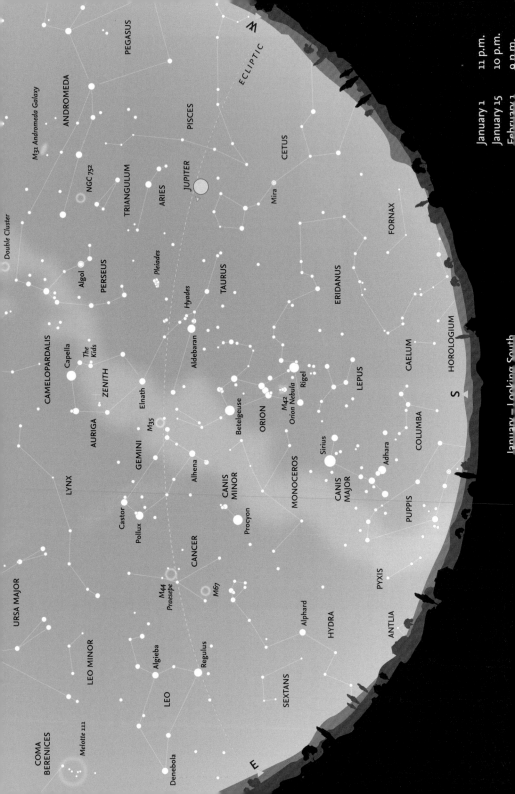

January – Looking South

URSA MAJOR
COMA BERENICES
Melotte 111
LEO MINOR
Algieba
LEO
Regulus
Denebola
CANCER
M44 Praesepe
M67
Castor
Pollux
Alphard
SEXTANS
HYDRA
PYXIS
ANTLIA
PUPPIS
CANIS MAJOR
Adhara
Sirius
Procyon
CANIS MINOR
Alhena
MONOCEROS
GEMINI
M35
Elnath
ORION
Betelgeuse
M42 Orion Nebula
Rigel
LEPUS
COLUMBA
CAELUM
HOROLOGIUM
ERIDANUS
Aldebaran
Hyades
TAURUS
FORNAX
CETUS
Mira
AURIGA
ZENITH
Capella
The Kids
CAMELOPARDALIS
LYNX
PERSEUS
Algol
Pleiades
Double Cluster
NGC 752
TRIANGULUM
ARIES
JUPITER
ECLIPTIC
PISCES
ANDROMEDA
M31 Andromeda Galaxy
PEGASUS

S
E

January 1   11 p.m.
January 15   10 p.m.
February 1   9 p.m.

January – Looking South

The southern sky is dominated by **Orion**, prominent during the winter months, visible at some time during the night. It is highly distinctive, with a line of three stars that form the "Belt." To most observers, the bright star **Betelgeuse** (α Orionis), shows a reddish tinge, in contrast to the brilliant bluish-white **Rigel** (β Orionis). The three stars of the belt lie directly south of the celestial equator. A vertical line of three "stars" forms the "Sword" that hangs south of the Belt. With good viewing, the central "star" appears as a hazy spot, even to the naked eye, and is actually the **Orion Nebula**. Binoculars reveal the four stars of the Trapezium, which illuminate the nebula.

Orion's Belt points up to the northwest toward **Taurus** and orange-tinted **Aldebaran** (α Tauri). Close to Aldebaran, there is a conspicuous "V" of stars, called the **Hyades** cluster. (Despite appearances, Aldebaran is not part of the cluster.) Farther along, the same line from Orion passes below a bright cluster of stars, the **Pleiades**, or Seven Sisters. Even the smallest pair of binoculars reveals this as a beautiful group of bluish-white stars. The two most conspicuous of the other stars in Taurus lie directly above Orion, and form an elongated triangle with Aldebaran. The northernmost, **Elnath** (α Tauri), was once considered to be part of the constellation of **Auriga**.

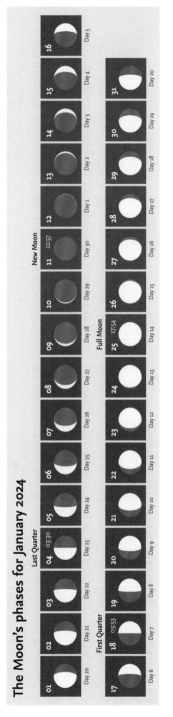

The constellation of Orion dominates the sky during this period of the year, and is a useful starting point for recognizing other constellations in the southern sky. Here, orange Betelgeuse, blue-white Rigel and the pinkish Orion Nebula are prominent. Orion can be found in the southern part of the sky (see page 36).

The Moon's phases for January 2024

Last Quarter

01	02	03	04 03:30	05	06	07	08	09	10	11 11:57	12	13	14	15	16
Day 20	Day 21	Day 22	Day 23	Day 24	Day 25	Day 26	Day 27	Day 28	Day 29	Day 30	Day 1	Day 2	Day 3	Day 4	Day 5

First Quarter · New Moon

17	18 03:53	19	20	21	22	23	24	25 17:54	26	27	28	29	30	31
Day 6	Day 7	Day 8	Day 9	Day 10	Day 11	Day 12	Day 13	Day 14	Day 15	Day 16	Day 17	Day 18	Day 19	Day 20

Full Moon

January – Moon and Planets

The Earth

The Earth reaches perihelion (the closest point to the Sun in its yearly orbit) on January 3 at 03:39 Universal Time, when it is at a distance of 0.98306994 AU (147,101,081 km).

The Moon

On January 4, at Last Quarter, the Moon is 2°N of **Spica** in **Virgo**. On January 8 it passes 0.8°N of **Antares**, and a few hours later, is 5.7°S of **Venus**. It passes a similar distance south of **Mercury** the next day, and 4.2°S of **Mars** on January 10. On January 14, the waxing crescent Moon is 2.1°S of **Saturn**, and just 1.0°S of **Neptune** the next day. At First Quarter, the Moon is 1.8°N of **Jupiter** and 3.0°N of much fainter **Uranus** the following day. It then passes 9.5°N of **Aldebaran** on January 21.

By January 24, a day before Full Moon, it is 1.7°S of **Pollux**. On January 27 the Moon passes 3.9°N of **Regulus**, between it and **Algieba**.

The planets

Mercury comes to western elongation on January 12 (see the diagram on page 22). **Venus** is bright (mag. -4.0), not far from Mercury in the morning sky. **Mars** (mag. 1.3 to 1.4) is in the same area, but lower in the sky. **Jupiter** (mag. -2.4 to -2.5) is in southern **Aries** and **Saturn** (mag. 1.0) is in **Aquarius**. **Uranus** (mag. 5.7) is in **Aries** and **Neptune** (mag. 7.9) in **Pisces**.

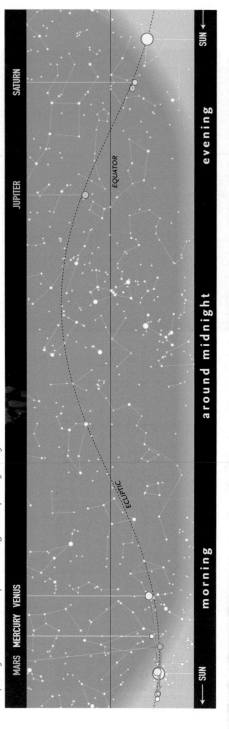

The path of the Sun and the planets along the ecliptic in January.

Calendar for January

Early morning 3 a.m.

January 4–5 • Shortly after Last Quarter, the Moon passes Spica, near the southeast.

Morning 6:45 a.m.

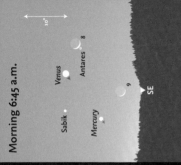

January 8–9 • On Janauy 8, the Moon is close to Antares. Later it passes Venus and Mercury. Sabik is nearby.

Evening 6 p.m.

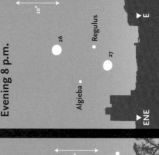

January 13–14 • The crescent Moon passes Saturn. Fomalhaut and Saturn are equally bright.

Evening 11 p.m.

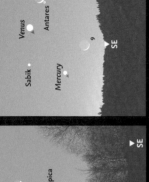

January 18–20 • The Moon passes Jupiter, Hamal, the Pleiades and Aldebaran. Menkar (α Cet) is nearby.

Evening 6 p.m.

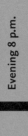

January 24 • The Full Moon lines up with Pollux and Castor. Alhena and Procyon are almost due east.

Evening 8 p.m.

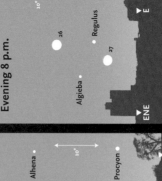

January 26–27 • A few days after Full Moon, it passes between Regulus and Algieba.

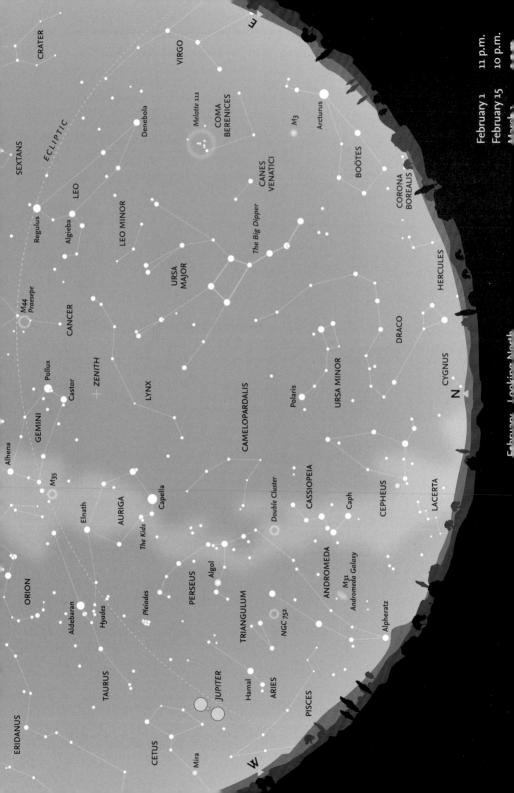

February, Looking North

February 1   11 p.m.
February 15   10 p.m.
March 1

February – Looking North

The months of January and February are probably the best time for seeing the section of the Milky Way that runs in the northern and western sky from **Cygnus**, low on the northern horizon, through **Cassiopeia**, **Perseus** and **Auriga** and then down through **Gemini** and **Monoceros** (see chart on page 42). Although not as readily visible as the denser star clouds of the summer Milky Way, on a clear night so many stars may be seen that even a distinctive constellation such as Cassiopeia is not immediately obvious.

The head of **Draco** is now higher in the sky and easier to recognize. **Deneb** (α Cygni), the brightest star in **Cygnus**, is hidden below the horizon, almost due north at midnight. **Vega** (α Lyrae) in **Lyra** is so low that it is difficult to see, but may become visible later in the night. The constellation of **Boötes** – sometimes described as shaped like a kite, an ice-cream cone, or the letter "P" – with orange-tinted **Arcturus** (α Boötis), is beginning to clear the eastern horizon. Arcturus, at magnitude -0.05, is the brightest star in the northern hemisphere of the sky. The inconspicuous constellation of **Coma Berenices** is now well above the horizon in the east. The concentration of faint stars at the northwestern corner somewhat resembles a tiny, detached portion of the Milky Way. This is Melotte 111, an open star cluster (which is sometimes called the Coma Cluster, but must not be confused with the important Coma Cluster of galaxies, Abell 1656, mentioned on page 55).

On the other side of the sky, in the northwest, most of the constellation of **Andromeda** is still easily seen, although **Alpheratz** (α Andromedae), the star that forms the northeastern corner of the Great Square of Pegasus – even though it is actually part of Andromeda – is becoming close to the horizon and more difficult to detect. High overhead, at the zenith, try to make out the very faint constellation of

Lynx. It was introduced in 1687 by the famous astronomer Johannes Hevelius to fill the largely blank area between **Auriga**, **Gemini** and **Ursa Major**, and is reputed to be so named because one needed the eyes of a lynx to detect it.

A very large, and frequently ignored, open star cluster, Melotte 111, also known as the Coma Cluster, is readily visible in the eastern sky during February.

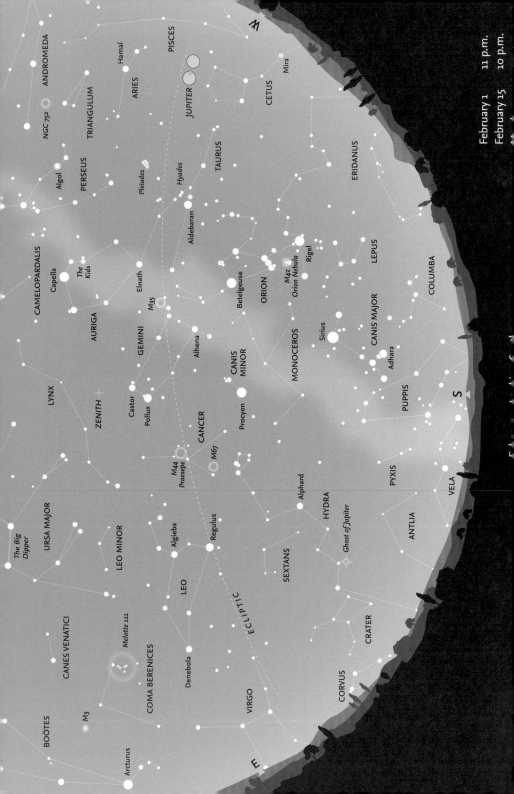

February 1   11 p.m.
February 15   10 p.m.

February – Looking South

Apart from **Orion**, the most prominent constellation visible this month is **Gemini**, with its two lines of stars running southwest toward **Orion**. Many people have difficulty in remembering which is which of the two stars **Castor** and **Pollux**. Castor (α Geminorum), the fainter star (mag. 1.9), is closer to the North Celestial Pole. Pollux (β Geminorum) is the brighter of the two (mag. 1.2), but is farther away from the Pole. Pollux is one of the first-magnitude stars that may be occulted by the Moon. Castor is remarkable because it is actually a multiple system, consisting of no less than six individual stars.

Orion's belt points down to the southeast toward **Sirius**, the brightest star in the sky (at magnitude -1.4) in the constellation of **Canis Major**, the whole of which is now clear of the southern horizon. Forming an equilateral triangle with **Betelgeuse** in Orion and Sirius in Canis Major is **Procyon**, the brightest star in the small constellation of **Canis Minor**. Between Canis Major and Canis Minor is the faint constellation of **Monoceros**, which actually straddles the Milky Way, which, although faint, has many clusters in this area. Directly east of Procyon is the highly distinctive asterism of six stars that form the "head" of **Hydra**,

The constellation of Gemini. The two brightest stars, visible near the left-hand top corner of the photograph, are Castor (top) and Pollux.

the largest of all 88 constellations, and which trails such a long way across the sky that it is only in mid-March around midnight that the whole constellation becomes visible.

The Moon's phases for February 2024

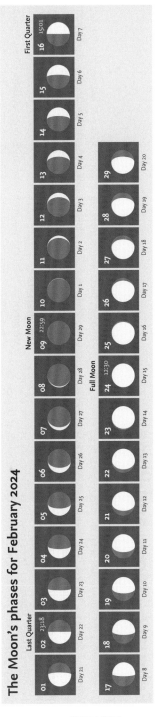

					Last Quarter		
01	**02** 23:18	**03**	**04**	**05**	**06**	**07**	**08**
Day 21	Day 22	Day 23	Day 24	Day 25	Day 26	Day 27	Day 28

New Moon							First Quarter
09 22:59	**10**	**11**	**12**	**13**	**14**	**15**	**16** 15:01
Day 29	Day 1	Day 2	Day 3	Day 4	Day 5	Day 6	Day 7

Full Moon							
17	**18**	**19**	**20**	**21**	**22**	**23**	**24** 12:30
Day 8	Day 9	Day 10	Day 11	Day 12	Day 13	Day 14	Day 15

25	**26**	**27**	**28**	**29**			
Day 16	Day 17	Day 18	Day 19	Day 20			

February – Moon and Planets

The Moon

On February 1 the Moon, just before Last Quarter, is 1.7°N of *Spica* in *Virgo*. By February 5 it is 0.6°N of orange-red *Antares*. One day later it is 5.4°S of brilliant *Venus*. It passes 4.2°S of *Mars* the next day and 3.2°S of *Mercury* later that day. Two days after New Moon (by February 11) the Moon is 1.8°S of *Saturn* and the next day is even closer (0.7°S) to the much fainter *Neptune*. Three days later (February 15) the Moon is 3.2°N of *Jupiter* and the next day the same distance north of *Uranus*. On February 17, one day after First Quarter, the Moon is 9.8°N of *Aldebaran* in *Taurus*. By February 21, it is 1.6°S of *Pollux*, three days before Full Moon. On February 23, the Moon is 3.6°N of *Regulus*, passing between it and *Algieba*.

Occultations

Although there are as many as 14 occultations of *Antares* in 2024 and 9 of *Spica*, none are readily visible from land, visible (if at all) from the oceans. There are no occultations in 2024 of the other bright stars (*Aldebaran*, *Pollux* and *Regulus*) near the ecliptic.

The planets

Mercury is 3.2°N of the Moon on February 8, one day before New Moon. *Venus* remains bright (mag. -3.9) in the morning sky, but is becoming lower towards the horizon. *Mars* is too close to the Sun to be readily visible. *Jupiter* is moving slowly in *Aries* at mag. -2.3 to -2.2. *Saturn* (mag. 1.0) remains in *Aquarius*. *Uranus* (mag. 5.7 to 5.8) is in *Aries* and *Neptune* (mag. 7.9 to 8.0) in *Pisces*.

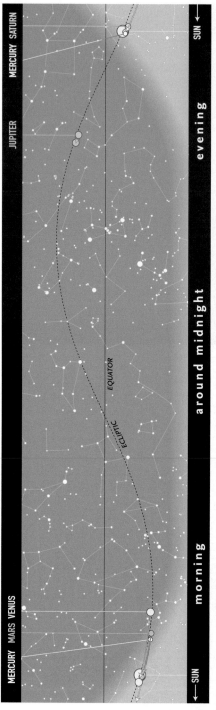

The path of the Sun and the planets along the ecliptic in February.

Calendar for February

01	07:46	Spica 1.7°S of the Moon
02	23:18	Last Quarter
05	00:52	Antares 0.6°S of the Moon
07	18:50	Venus (mag. -3.9) 5.4°N of the Moon
08	06:30	Mars (mag. 1.3) 4.2°N of the Moon
08	21:59	Mercury (mag. -0.4) 3.2°N of the Moon
09	22:59	New Moon
10	18:53	Moon at perigee = 358,099 km
11	00:40	Saturn (mag. 1.0) 1.8°N of the Moon
12	06:45	Neptune (mag. 7.9) 0.7°N of the Moon
15	08:16	Jupiter (mag. -2.3) 3.2°S of the Moon
16	08:16	Uranus (mag. 5.8) 3.2°S of the Moon
16	15:01	First Quarter
17	16:41	Aldebaran 9.8°S of the Moon
21	01:33	Pollux 1.6°N of the Moon
23	23:26	Regulus 3.6°S of the Moon
24	12:30	Full Moon
25	14:59	Moon at apogee = 406,312 km
28	14:00 *	Saturn (mag. 0.9) 0.2°N of Mercury (mag. -1.8)
28	14:23	Spica 1.5°S of the Moon

These objects are close together for an extended period around this time.

After midnight 0:30 a.m.

February 1 • When the Moon rises, in the east-southeast, it is close to Spica.

Morning 6 a.m.

February 4–5 • The Moon passes Antares. The Cat's Eyes (λ and υ Sco) are close to the horizon.

Morning 6:30 a.m.

February 7 • Shortly before sunrise the narrow crescent Moon is in the company of Venus and Mars.

Evening 10 p.m.

February 14–16 • The Moon passes Jupiter, Hamal, the Pleiades and Aldebaran, in the western sky.

Evening 9 p.m.

February 20 • Very high in the southeast, the Moon lines up with Pollux and Castor.

Evening 6 p.m.

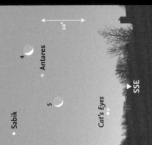

February 23 • When the Moon rises it is between Regulus and Algieba.

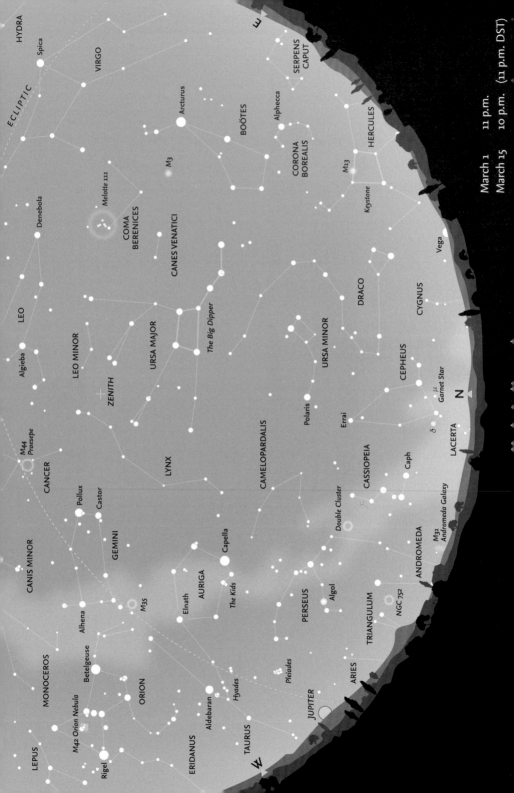

March 1   11 p.m.
March 15 10 p.m. (11 p.m. DST)

March – Looking North

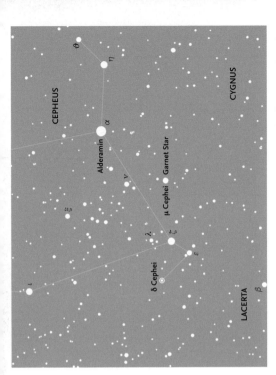

A finder chart for δ Cephei and μ Cephei (the Garnet Star). All stars brighter than magnitude 7.5 are shown. A photograph of the area is on page 49.

The Sun crosses the celestial equator on Tuesday, March 19, at the vernal equinox, when day and night are of almost equal length, and the northern season of spring is considered to begin. (The hours of daylight and darkness change most rapidly around the equinoxes in March and September.) It is also in March that Daylight Saving Time (DST) begins (on Saturday–Sunday, March 9–10) so the charts show the appearance at 11 p.m. for March 1 and 10 p.m. (DST) for April 1. (In Europe, Daylight Saving Time is introduced three weeks later.)

Early in the month, the constellation of **Cepheus** lies almost due north, with the distinctive "W" of **Cassiopeia** to its west. Cepheus lies across the border of the Milky Way and is often described as like the gable-end of a house and is best seen later in the night or in the month. Despite the large number of stars revealed at the base of the constellation by binoculars, one star stands out because of its deep red color. This is **μ Cephei**, also known as the Garnet Star, because of its striking color. It is a truly gigantic star, a red supergiant, and one of the largest stars known. It is about 1,400 times the diameter of the Sun, and if placed in the Solar System would extend beyond the orbit of Jupiter. (Betelgeuse, in Orion, is also a red supergiant, but it is "only" about 500 times the diameter of the Sun.)

Another famous, and very important star in Cepheus is **δ Cephei**, which is the prototype for the class of variable stars known as Cepheids. These giant stars show a regular variation in their luminosity, and there is a direct relationship between the period of the changes in magnitude and the stars' actual luminosity. From a knowledge of the period of any Cepheid, its actual luminosity – known as its absolute magnitude – may be derived. A comparison of its apparent magnitude on the sky and its absolute magnitude enables the star's exact distance to be determined. Once the distances to the first Cepheid variables had been established,

examples in more distant galaxies provided information about the scale of the universe. Cepheid variables are the first major "rung" in the cosmic distance ladder. Both important stars are shown on the accompanying chart and in the photograph on page 49.

Below Cepheus to the east (to the right), late in the night, it may be possible to catch a glimpse of **Deneb** (α Cygni), just above the horizon. Slightly farther round toward the northeast, **Vega** (α Lyrae) is marginally higher in the sky. From most of the United States, Deneb is just far enough north to be seen at some time during the night (although difficult to see in January and February because it is so low). Vega, by contrast, farther south, is completely hidden during the depths of winter.

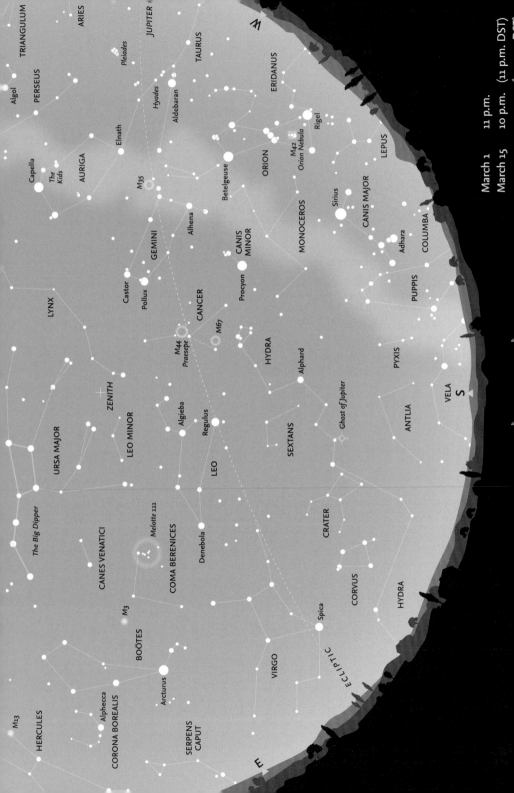

March 1   11 p.m.
March 15   10 p.m. (11 p.m. DST)

March – Looking South

Due south at 10 p.m. at the beginning of the month, lying between the constellations of **Gemini** in the west and **Leo** in the east, and fairly high in the sky above the head of Hydra, is the faint, and rather undistinguished zodiacal constellation of **Cancer**. Rather like an upside-down letter "Y," it has three "legs" radiating from the center, where there is an open cluster, M44 or **Praesepe** ("the Manger," but also known as "the Beehive"). On a clear night this cluster, known since antiquity, is just a hazy spot to the naked eye, but appears in binoculars as a group of dozens of individual stars.

Also prominent in March is the constellation of Leo, with the "backward question mark" (or "Sickle") of bright stars forming the head of the mythological lion. **Regulus** (α Leonis) – the "dot" of the "question mark" or the handle of the sickle and the brightest star in Leo – lies very close to the ecliptic and is one of the few first-magnitude stars that may be occulted by the Moon. However, there are no occultations of Regulus in 2024, nor of any of the other four bright stars near the ecliptic.

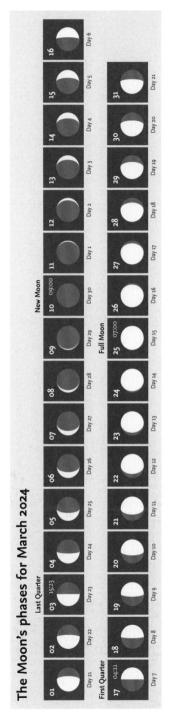

The constellation of Cepheus, with the "W" of Cassiopeia on the left. The locations of both δ and μ Cephei are shown on the chart on page 47.

The Moon's phases for March 2024

01	02	**03** 15:23	04	05	06	07	08	09	**10** 09:00	11	12	13	14	15	16
Day 21	Day 22	Day 23	Day 24	Day 25	Day 26	Day 27	Day 28	Day 29	Day 30	Day 1	Day 2	Day 3	Day 4	Day 5	Day 6

First Quarter (01), Last Quarter (03), New Moon (10)

17 04:11	18	19	20	21	22	23	24	**25** 07:00	26	27	28	29	30	**31**
Day 7	Day 8	Day 9	Day 10	Day 11	Day 12	Day 13	Day 14	Day 15	Day 16	Day 17	Day 18	Day 19	Day 20	Day 21

First Quarter (17), Full Moon (25)

March – Moon and Planets

The Moon

On March 3, at Last Quarter, the Moon is 0.4°N of **Antares**. On March 8 it passes 3.5°S of **Mars**. Later the same day it is 3.3°S of **Venus**. The next day (March 9) it is 1.5°S of **Saturn** (mag.1.0). On March 11, a few hours after New Moon, it is 0.5°S of **Neptune** (mag.8.0) and the next day it is 1.0°S of **Mercury** (mag.-1.3). On March 14 it is 4.0°N of **Jupiter** (mag.-2.1) and later that day 3.4°N of **Uranus** (mag. 5.8). On March 15 it is 9.9°S of **Aldebaran** in **Taurus**. On March 19, just after First Quarter, it is 1.4°S of **Pollux**. By March 22 it is 3.6°N of **Regulus**. At Full Moon (March 25) there is a penumbral lunar eclipse (visible from the USA) when the Moon may just touch the umbra. The next day the Moon is 1.4°N of **Spica**. On March 30 the Moon is 0.3°N of **Antares**.

The Planets

Mercury (mag.-0.2) reaches greatest eastern elongation on March 24 (see the diagram on page 22). **Venus** is bright (mag.-3.8) but is closing rapidly on the Sun, and may be glimpsed early in the month. **Mars** (mag.-1.3 to 1.2) is similarly placed. **Jupiter** (mag.-2.2 to -2.1) is in **Aries** and **Saturn** (mag.1.0 to 1.1) remains in **Aquarius**. **Uranus** (mag.5.8) is in **Aries** and **Neptune** (mag. 8.0) in **Pisces**.

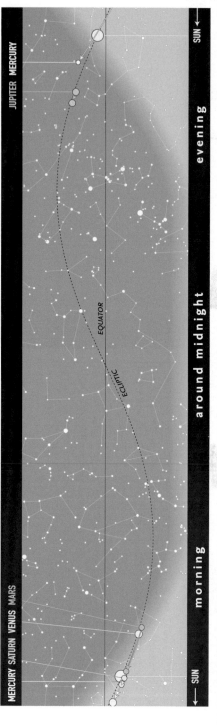

The path of the Sun and the planets along the ecliptic in March.

Early morning 3 a.m.

Sabik • Antares

Moon

SE ▶

March 3 • *The Last Quarter Moon is close to Antares. Sabik is farther east.*

Morning 6 a.m.

Moon

Mars

Venus

ESE   SE

March 7 • *In the morning twilight, the Moon, Venus and Mars (mag. 1.2) are low in the southeast.*

Evening 9 p.m. (DST)

10°

Pleiades   15

14

Aldebaran   13

Jupiter

W   40°

March 13–15 • *High in the west, the Moon passes Jupiter, the Pleiades and Aldebaran.*

Early morning 3 a.m. (DST)

35°   Algieba

22   Regulus

21

20

Castor

Pollux   19

WNW   10°   W ▶

WSW ▶

March 19–22 • *Within a few days, the waxing gibbous Moon moves from Pollux and Castor, to Regulus and Algieba in the western sky.*

Morning 5 a.m. (DST)

10°

Sabik   Moon

Antares

25°

Cat's Eyes

S

March 30 • *The Moon is near Antares, with Sabik and the Cat's Eyes farther east.*

Calendar for March

03	08:54	Antares 0.4°S of the Moon
03	15:23	Last Quarter
03	18:01	Minor planet (3) Juno at opposition (mag. 8.7)
08	05:00	Mars (mag. 1.2) 3.5°N of the Moon
08	08:18 *	Neptune (mag. 8.0) 0.5°S of Mercury (mag. –1.5)
08	17:00	Venus (mag. –3.8) 3.3°N of the Moon
09	17:28	Saturn (mag. 1.0) 1.5°N of the Moon
10		Summer Time begins
10	07:04	Moon at perigee = 356,896 km (closest of the year)
10	19:26	Neptune (mag. 8.0) 0.5°N of the Moon
11	02:31	Mercury (mag. –1.3) 1.0°N of the Moon
11	09:00	New Moon
14	01:02	Jupiter (mag. –2.1) 4.0°S of the Moon
14	11:35	Uranus (mag. 5.8) 3.4°S of the Moon
15	23:43	Aldebaran 9.9°N of the Moon
17	04:11	First Quarter
19	07:24	Pollux 1.4°N of the Moon
22	05:27	Regulus 3.6°S of the Moon
23	15:45	Moon at apogee = 406,294 km
24	22:34	Mercury at greatest elongation (18.7°E, mag. –0.2)
25	07:00	Full Moon
25	07:14	Penumbral lunar eclipse
26	20:23	Spica 1.4°S of the Moon
30	15:03	Antares 0.3°S of the Moon

* *These objects are close together for an extended period around this time.*

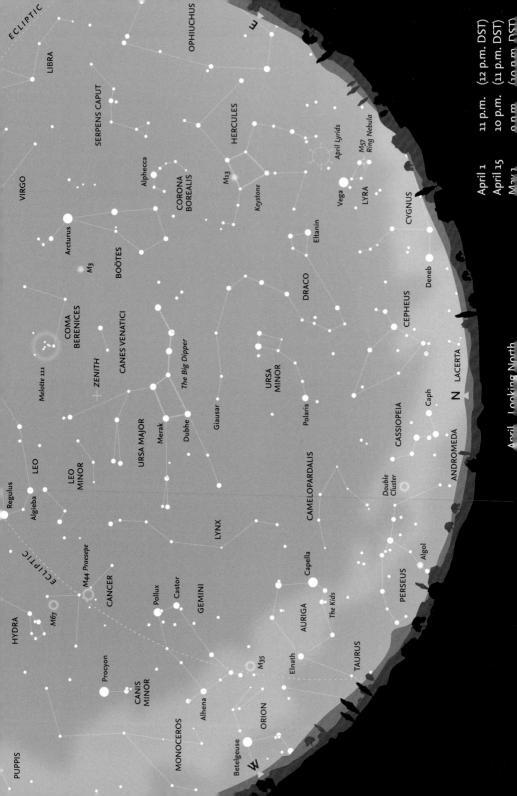

ECLIPTIC

OPHIUCHUS

LIBRA

E

SERPENS CAPUT

HERCULES

VIRGO

Alphecca

April Lyrids

CORONA
BOREALIS

M57
Ring Nebula

M13

Arcturus

Vega

CYGNUS

M3

Keystone

LYRA

BOÖTES

Eltanin

DRACO

Deneb

COMA
BERENICES

CEPHEUS

CANES VENATICI

Melotte 111

The Big Dipper

URSA
MINOR

ZENITH

Giausar

Polaris

LACERTA

URSA MAJOR

N

Merak

Dubhe

CASSIOPEIA

Caph

LEO

Algieba

LEO
MINOR

CAMELOPARDALIS

ANDROMEDA

Regulus

Double
Cluster

LYNX

Algol

ECLIPTIC

Capella

PERSEUS

M44 Praesepe

CANCER

Pollux

The Kids

HYDRA

M67

Castor

AURIGA

TAURUS

GEMINI

Procyon

CANIS
MINOR

Elnath

M35

Alhena

ORION

Betelgeuse

MONOCEROS

W

PUPPIS

April, Looking North

April 1   11 p.m.   (12 p.m. DST)
April 15   10 p.m.   (11 p.m. DST)
May 1   9 p.m.   (10 p.m. DST)

April – Looking North

Cygnus and the brighter regions of the Milky Way are now starting to become visible later in the night. Rising in the northeast is the small constellation of **Lyra** and the distinctive "Keystone" of **Hercules** above it. This asterism is very useful for locating the bright globular cluster M13 (see map on page 59), which lies on one side of the quadrilateral. The winding constellation of **Draco** weaves its way from the four stars that marks its "head," on the border with Hercules, to end at λ Draconis between **Polaris** (α Ursae Minoris) and the "Pointers," **Dubhe** and **Merak** (α and β Ursae Majoris, respectively). **Ursa Major** is "upside down" high overhead, near the zenith. The constellation of **Gemini** stands almost vertically in the west. **Auriga** is still clearly seen in the northwest, but, by the end of the month, the southern portion of **Perseus** is starting to dip below the northern horizon. The very faint constellation of **Camelopardalis** lies in the northwest between Polaris and the constellations of Auriga and Perseus.

Meteors

There is one moderate meteor shower this month, the **Lyrids** (often known as the **April Lyrids**). This year the shower begins on April 14, one day before First Quarter, but peaks on April 22 to 23 at Full Moon, so begins reasonably favorably but becomes very unfavorable. There is another, stronger shower, the **η-Aquariids**, that begins on April 19, when the Moon is waxing gibbous, and continues until May 28, well after the next New Moon. The shower is best seen in early May.

The constellation of Hercules (center), with Lyra (bottom left) and Corona Borealis (top right).

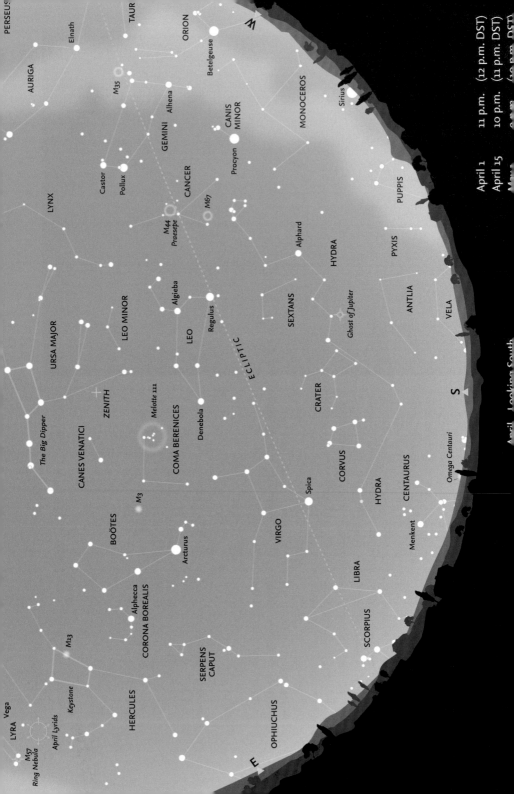

April. Looking South

PERSEUS

TAUR

ORION

AURIGA

Elnath

M35

Betelgeuse

Allhena

Sirius

GEMINI

CANIS
MINOR

MONOCEROS

Castor

Pollux

CANCER

Procyon

LYNX

M44
Praesepe

M67

PUPPIS

Algieba

Regulus

Alphard

HYDRA

PYXIS

URSA MAJOR

LEO MINOR

LEO

SEXTANS

ANTLIA

VELA

Ghost of Jupiter

ZENITH

Melotte 111

The Big Dipper

CANES VENATICI

COMA BERENICES

Denebola

ECLIPTIC

CRATER

S

M3

Spica

CORVUS

CENTAURUS

BOÖTES

VIRGO

HYDRA

Omega Centauri

Arcturus

Menkent

Alphecca

LIBRA

CORONA BOREALIS

M13

SERPENS
CAPUT

SCORPIUS

Keystone

HERCULES

Vega

LYRA

April Lyrids

OPHIUCHUS

M57
Ring Nebula

E

April – Looking South

Leo is the most prominent constellation in the southern sky in April, and vaguely looks like the creature after which it is named. *Gemini*, with *Castor* and *Pollux*, remains clearly visible in the west, and *Cancer* lies between the two constellations. To the east of Leo, the whole of *Virgo*, with *Spica* (α Virginis) its brightest star, is well clear of the horizon and *Libra* is coming into view. Below Leo and Virgo, the complete length of *Hydra* is visible, running beneath both constellations, with *Alphard* (α Hydrae) halfway between *Regulus* and the southwestern horizon. Farther east, the two small constellations of *Crater* and the rather brighter *Corvus* lie between Hydra and Virgo. Part of *Centaurus* is now above the horizon.

Boötes and *Arcturus* are prominent in the eastern sky, together with the circlet of *Corona Borealis*, framed by Boötes and the neighboring constellation of *Hercules*. Between Leo and Boötes lies the constellation of *Coma Berenices*, notable for being the location of the open cluster Melotte 111 (see page 41) and the Coma Cluster of galaxies (Abell 1656). There are about 1,000 galaxies in this cluster, which is located near the North Galactic Pole, where we are looking out of the plane of the

The distinctive constellation of Leo, with Regulus and "The Sickle" on the west. Algieba (γ Leonis), north of Regulus, appearing double, is a multiple system of four stars.

Galaxy and are thus able to see deep into space. Only about 10 of the brightest galaxies in the Coma Cluster are visible with the largest amateur telescopes.

The Moon's phases for April 2024

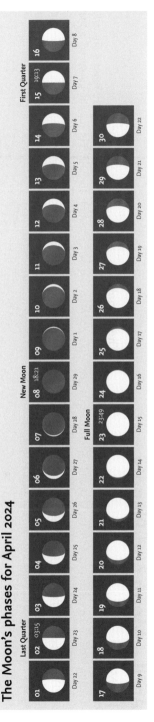

April – Moon and Planets

The Moon

On April 8, at New Moon, there is a total solar eclipse, with the track covering Mexico, the eastern states of the USA and eastern Canada (see pages 20 and 21). The Moon then passes the **Pleiades** and by April 12 is 10°N of **Aldebaran** in **Taurus**. It is 1.5°S of **Pollux** by April 15 and 3.5°N of **Regulus**, between it and **Algieba** on April 18. Before Full Moon on April 23, it is 1.4°N of **Spica** in **Virgo** and 0.3°N of **Antares** by April 26.

The planets

Mercury is too close to the Sun to be visible this month. **Venus** is similarly placed. **Mars** is moving towards the Sun, but may be glimpsed in the morning sky early in the month. **Jupiter** (mag. -2.0) is in **Aries**, crossing into **Taurus** at the end of the month. **Saturn** (mag. 1.1 to 1.2) is moving slowly in **Aquarius**. **Uranus** (mag. 5.8) remains in **Aries** and **Neptune** (mag. 8.0 to 7.9) in **Pisces**.

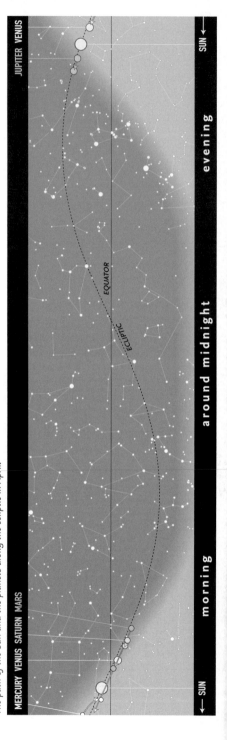

The path of the Sun and the planets along the ecliptic in April.

Calendar for April

02	03:15	Last Quarter
06	03:51	Mars (mag. 1.2) 2.0°N of the Moon
06	09:24	Saturn (mag. 1.1) 1.2°N of the Moon
07	08:11	Neptune (mag. 8.0) 0.4°N of the Moon
07	16:38	Venus (mag. -3.8) 0.4°S of the Moon
07	17:51	Moon at perigee = 358,849 km
08	18:17	Total solar eclipse (Mexico/USA)
08	18:21	New Moon
09	01:24	Mercury (mag. 4.9) 2.2°N of the Moon
10	21:09	Jupiter (mag. -2.0) 4.0°S of the Moon
10	23:51	Uranus (mag. 5.8) 3.6°S of the Moon
12	08:48	Aldebaran 10°S of the Moon
14–30		April Lyrid meteor shower
15	14:25	Pollux 1.5°N of the Moon
15	19:13	First Quarter
18	11:56	Regulus 3.5°S of the Moon
18	23:00 *	Venus (mag. -3.8) 2.0°S of Mercury (mag. 3.9)
19–May 28		η-Aquariid meteor shower
20	02:10	Moon at apogee = 405,623 km
22–23		April Lyrid meteor shower maximum
23	02:44	Spica 1.4°S of the Moon
23	23:49	Full Moon
26	20:39	Antares 0.3°S of the Moon

* These objects are close together for an extended period around this time.

Evening 9 p.m. (DST)

Aldebaran • Algol •
Pleiades
11
Jupiter
10
W ▽   WNW ▽
10°

April 10–11 • The narrow crescent Moon passes Jupiter, the Pleiades and Aldebaran. Algol (β Per) is farther north.

Evening 10 p.m. (DST)

Spica
Moon
30°
SE ▽

April 22 • High in the southeast, the Moon and Spica are side by side.

Early morning 3 a.m. (DST)

Denebola •
• Algieba
19
18
Regulus •
W ▽   WNW ▽
10°

April 18–19 • The Moon passes between Regulus and Algieba, near the western horizon. Denebola (β Leo) is about fifteen degrees higher and a little farther south.

Early morning 3 a.m. (DST)

Antares
26
27
•• Cat's Eyes
S ▽
10°

April 26–27 • The Moon passes Antares. The Cat's Eyes are closer to the horizon.

Evening 8:30 p.m. (DST)

Aldebaran •
Pleiades
Jupiter
W ▽   WNW ▽
10°

April 27 • After sunset, Jupiter is eight degrees below the Pleiades. Aldebaran is farther west.

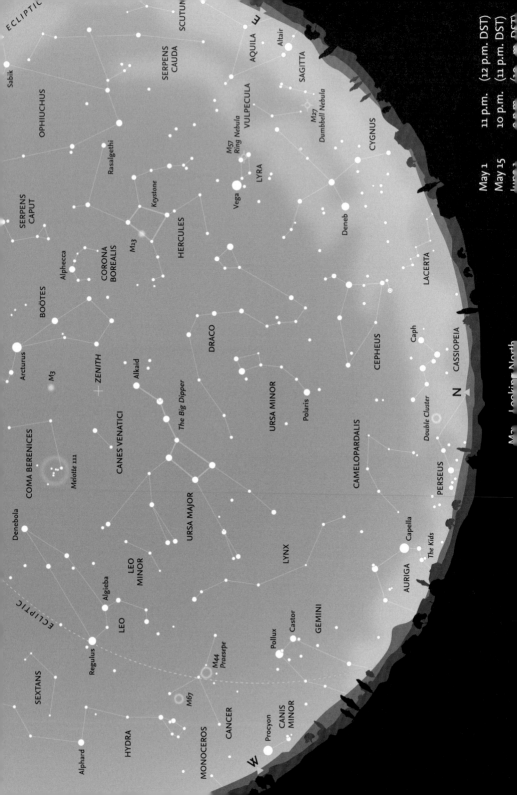

ECLIPTIC

SCUTUM

SERPENS
CAUDA

OPHIUCHUS

AQUILA

Altair

SAGITTA

Sabik

Rasalgethi

SERPENS
CAPUT

Keystone

M27
Dumbbell Nebula

M57
Ring Nebula

VULPECULA

Alphecca

CORONA
BOREALIS

Vega

LYRA

CYGNUS

M13

HERCULES

Deneb

BOÖTES

Arcturus

DRACO

LACERTA

M3

ZENITH

Alkaid

Caph

CEPHEUS

CASSIOPEIA

COMA BERENICES

The Big Dipper

URSA MINOR

N

Denebola

Melotte 111

CANES VENATICI

Polaris

Double Cluster

PERSEUS

LEO
MINOR

URSA MAJOR

CAMELOPARDALIS

Algieba

Capella

LEO

LYNX

The Kids

AURIGA

Regulus

ECLIPTIC

M44
Praesepe

Pollux

Castor

GEMINI

SEXTANS

M67

CANCER

Procyon

CANIS
MINOR

HYDRA

MONOCEROS

Alphard

W

May 1   11 p.m.   (12 p.m. DST)
May 15   10 p.m.   (11 p.m. DST)

Looking North

May – Looking North

Cassiopeia is now low over the northern horizon and, to its west, the southern portions of both **Perseus** and **Auriga** have been lost below the horizon, although the whole of **Draco** is still clearly visible. The constellations of **Lyra**, **Cepheus**, **Ursa Minor** and the whole of **Draco** are well placed in the sky. **Gemini**, with **Castor** and **Pollux**, is sinking towards the western horizon. **Capella** (α Aurigae) and the asterism of **The Kids** are still just clear of the horizon.

In the east, two of the stars of the "Summer Triangle," **Vega** (α Lyrae) and **Deneb** (α Cygni), are clearly visible, and the third star, **Altair** in **Aquila**, is beginning to climb above the horizon. The whole of **Cygnus** is now visible. The sprawling constellation of **Hercules** is high in the east and the brightest globular cluster in the northern hemisphere, M13, is visible to the naked eye on the western side of the asterism known as the **Keystone**.

Three faint constellations may be identified before the lighter nights of summer make them difficult objects. Below Cepheus, low in the northeastern sky is the zig-zag constellation of **Lacerta**, while to the west, above Perseus and Auriga is **Camelopardalis** and, farther west, the line of faint stars forming **Lynx**.

Later in the night (and in the month) the westernmost stars of **Pegasus** begin to come into view, while the stars of **Andromeda** start to appear on the northeastern horizon. High overhead, **Alkaid** (η Ursae Majoris), the last star in the "tail" of the Great Bear, is close to the zenith, while the main body of the constellation has swung round into the western sky.

Meteors

The **η-Aquariids** are one of the two meteor showers associated with Comet 1P/Halley (the other being the Orionids, in October). The η-Aquariids are not particularly favorably placed for northern-hemisphere observers, because the radiant is near the celestial equator, near the 'Water Jar' in Aquarius, well below the horizon until late in the night (around dawn). However, meteors may still be seen in the eastern sky even when the radiant is below the horizon. There is a radiant map for the η-Aquariids on page 30.

Their maximum in 2024, on May 6, occurs two days before New Moon, so is very favorable. Maximum hourly rate is about 50–55 per hour and a large proportion (about 25 per cent) of the meteors leave persistent trains.

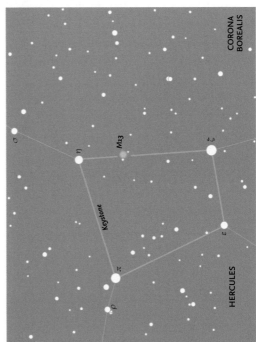

A finder chart for M13, the finest globular cluster in the northern sky. All stars down to magnitude 7.5 are shown.

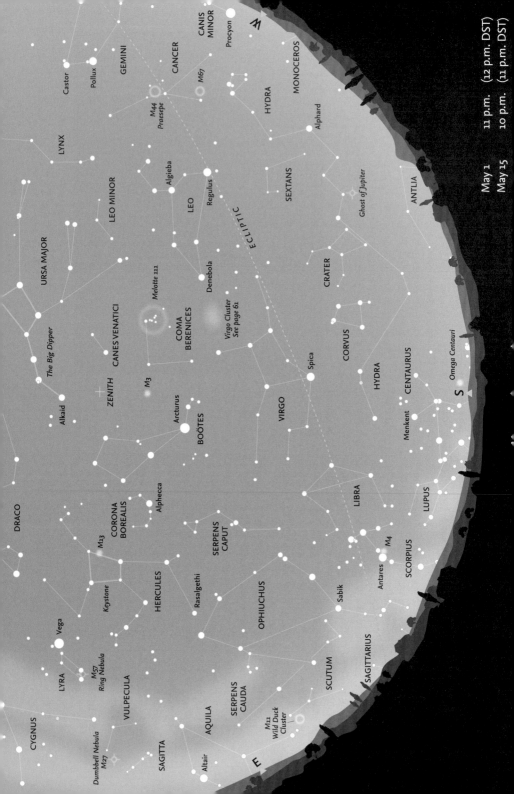

CANIS
MINOR

Procyon

CANCER

GEMINI

Castor

Pollux

M44
Praesepe

M67

HYDRA

Alphard

MONOCEROS

LYNX

Algieba

LEO

Regulus

SEXTANS

ANTLIA

Ghost of Jupiter

LEO MINOR

URSA MAJOR

Melotte 111

COMA
BERENICES

CRATER

ECLIPTIC

CANES VENATICI

Virgo Cluster
See page 61

The Big Dipper

ZENITH

M3

CORVUS

HYDRA

Spica

DRACO

Alkaid

Arcturus

VIRGO

CENTAURUS

Omega Centauri

S

BOÖTES

Menkent

CORONA
BOREALIS

Alphecca

M13

SERPENS
CAPUT

LIBRA

LUPUS

HERCULES

Rasalgethi

Keystone

Vega

OPHIUCHUS

Sabik

Antares   M4

SCORPIUS

LYRA

M57
Ring Nebula

VULPECULA

SCUTUM

SAGITTARIUS

SERPENS
CAUDA

CYGNUS

SAGITTA

AQUILA

M11
Wild Duck
Cluster

Dumbbell Nebula
M27

Altair

E

May – Looking South

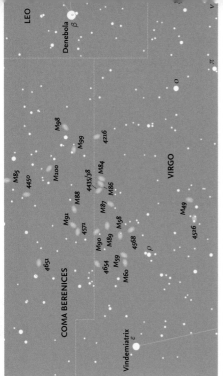

Early in the night, the constellation of **Virgo**, with **Spica** (α Virginis), lies due south, with **Leo** and both **Regulus** and **Denebola** (α and β Leonis, respectively) to its west still well clear of the horizon. The rather faint zodiacal constellation of **Libra** is now fully visible, together with most of **Scorpius** and ruddy **Antares** (α Scorpii). In the south, more of **Centaurus** may be seen, together with part of **Lupus**.

Virgo contains the nearest large cluster of galaxies, which is the center of the Local Supercluster, of which the Milky Way galaxy forms part. The Virgo Cluster contains some 2,000 galaxies, the brightest of which are visible in amateur telescopes.

Arcturus in **Boötes** is high in the south, with the distinctive circlet of **Corona Borealis** clearly visible to its east. The brightest star (α Coronae Borealis) is known as **Alphecca.** The large constellation of **Ophiuchus** (which actually crosses the ecliptic, and is thus the "13th" zodiacal constellation) is climbing into the eastern sky. Before the constellation boundaries were formally adopted by the International Astronomical Union in 1930, the southern region of Ophiuchus was regarded as

A finder chart for some of the brightest galaxies in the Virgo Cluster (see page 60). All stars brighter than magnitude 8.5 are shown.

forming part of the constellation of Scorpius, which had been part of the zodiac since antiquity.

The Moon's phases for May 2024

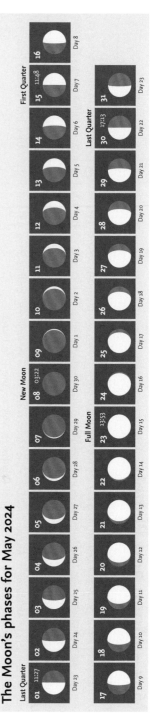

May – Moon and Planets

The Moon

On May 3, the Moon is 0.8°S of **Saturn** (mag. 1.2) and the next day is 0.3°S of much fainter **Neptune** (mag. 7.9). By May 5 it is 0.2°N of **Mars** (mag. 1.0). On May 6 it is 3.8°N of **Mercury** (mag. 0.6) and the next day 3.5°N of much brighter **Venus** (mag. -3.9). At New Moon on May 8, it is 3.6°N of faint **Uranus** (mag. 5.8) and a few hour later 4.3°N of **Jupiter** (mag. -2.0). On May 9 it is 9.9°N of **Aldebaran** in **Taurus**. On May 12 the Moon is 1.6°S of **Pollux**. At First Quarter on May 15, it is 3.5°N of **Regulus**. By May 20, the Moon is 1.4°N of **Spica** in **Virgo**. On May 24, the Moon (just past Full) is 0.4°N of **Antares**. By May 31, the Moon is 0.4°S of **Saturn.**

The planets

Mercury (mag. 0.4) reaches greatest elongation west in the evening sky on May 9. By May 31 (at mag. -0.7) it is 1.4°S of **Uranus** (mag. 5.8). **Venus** passes conjunction with the Sun, so is invisible this month. **Mars** is moving towards the Sun, so is also invisible. **Jupiter** (mag. -2.0), now in **Taurus**, is moving towards the **Hyades**. **Saturn** is moving slowly east in **Aquarius**. **Uranus** (mag. 5.8) remains in **Aries** and **Neptune** (mag. 7.9) in **Pisces**.

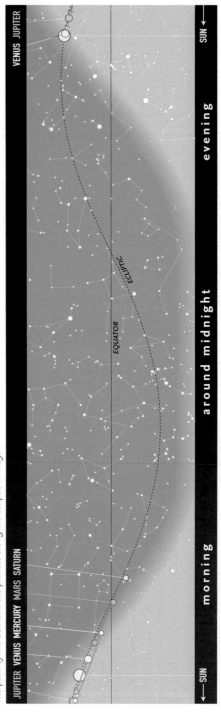

The path of the Sun and the planets along the ecliptic in May.

Calendar for May

01	11:27	Last Quarter
03	22:32	Saturn (mag. 1.2) 0.8°N of the Moon
04	18:55	Neptune (mag. 7.9) 0.3°N of the Moon
05	02:25	Mars (mag. 1.0) 0.2°S of the Moon
05	22:04	Moon at perigee = 363,163 km
06		η-Aquariid meteor shower maximum
06	08:25	Mercury (mag. 0.6) 3.8°S of the Moon
07	16:03	Venus (mag. -3.9) 3.5°S of the Moon
08	03:22	New Moon
08	12:51	Uranus (mag. 5.8) 3.6°S of the Moon
08	18:14	Jupiter (mag. -2.0) 4.3°S of the Moon
09	18:54	Aldebaran 9.9°S of the Moon
09	21:29	Mercury at greatest elongation (26.4°W, mag. 0.4)
12		Pollux 1.6°N of the Moon
15	11:48	First Quarter
15	19:24	Regulus 3.5°S of the Moon
17	18:59	Moon at apogee = 404,640 km
19	15:06	Minor planet (2) Pallas at opposition (mag. 9.0)
20	10:03	Spica 1.4°S of the Moon
23	13:53	Full Moon
24	03:10	Antares 0.4°S of the Moon
30	17:13	Last Quarter
31	01:00 *	Uranus (mag. 5.8) 1.4°N of Mercury (mag. -0.7)
31	08:09	Saturn (mag. 1.2) 0.4°N of the Moon

These objects are close together for an extended period around this time.

Morning 5:15 a.m. (DST)

May 3–5 • *The Moon passes Saturn, Mars and Mercury, which is probably lost in twilight.*

Evening 11 p.m. (DST)

May 12–15 • *The Moon is nearing First Quarter, when it moves from Pollux and Castor to Regulus and Algieba, high in the western sky.*

Morning 5 a.m. (DST)

May 31 • *The Moon meets Saturn, in the southeast. Fomalhaut is fifteen degrees lower.*

Early morning 3 a.m. (DST)

May 20 • *The Moon and Spica are side by side, about ten degrees above the horizon.*

After midnight 0:30 a.m. (DST)

May 24 • *The Moon is close to Antares. The Cat's Eyes are closer to the horizon.*

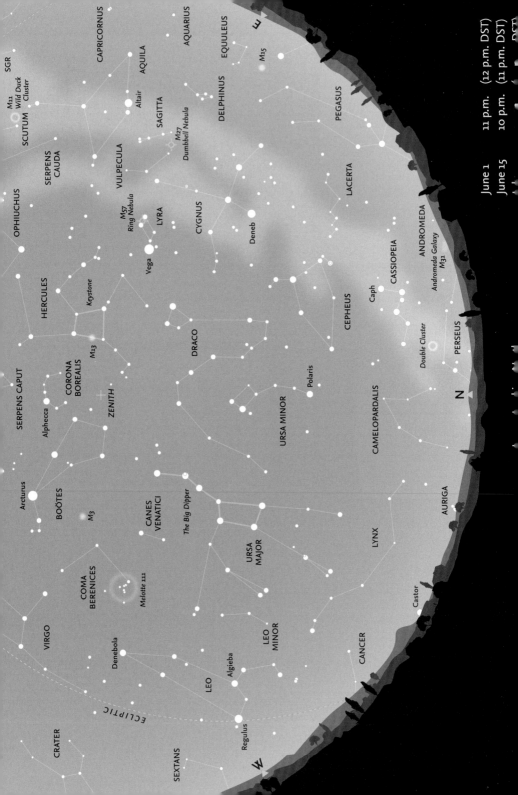

SGR
M11
Wild Duck
Cluster
SCUTUM
CAPRICORNUS
AQUILA
Altair
SAGITTA
AQUARIUS
EQUULEUS
M15
M27
Dumbbell Nebula
DELPHINUS
E
OPHIUCHUS
SERPENS
CAUDA
VULPECULA
M57
Ring Nebula
LYRA
CYGNUS
Deneb
PEGASUS
LACERTA
Vega
HERCULES
Keystone
DRACO
CEPHEUS
Caph
ANDROMEDA
CASSIOPEIA
Andromeda Galaxy
M31
M13
SERPENS CAPUT
CORONA
BOREALIS
ZENITH
Alphecca
URSA MINOR
Polaris
Double Cluster
PERSEUS
N
Arcturus
BOÖTES
M3
CANES
VENATICI
The Big Dipper
CAMELOPARDALIS
AURIGA
COMA
BERENICES
Melotte 111
URSA
MAJOR
LYNX
Castor
VIRGO
LEO
MINOR
CANCER
Denebola
Algieba
LEO
ECLIPTIC
Regulus
CRATER
SEXTANS
W

June 1   11 p.m.   (12 p.m. DST)
June 15   10 p.m.   (11 p.m. DST)

June – Looking North

With the approach of the summer solstice (June 20), twilight tends to persist in the northern United States and Canada, with four hours of darkness in Toronto in June, two hours in Seattle, and none in Vancouver. Even brighter stars, such as the seven stars making up the well-known asterism known as the **Big Dipper** in **Ursa Major** may be difficult to detect except around local midnight (1 a.m. DST). In the south of the United States, at New Orleans, Houston or Los Angeles, there are six or even seven hours of full darkness. Northern observers may have the compensation of seeing noctilucent clouds (NLC), electric-blue clouds visible during summer nights in the direction of the North Pole, for about a month or six weeks on either side of the solstice.

Two faint constellations, **Camelopardalis** and **Lynx**, may be glimpsed low on the northern and western horizons. **Leo** and **Regulus** (α Leonis) are heading downwards in the western sky, but **Cassiopeia**, farther east, is low, but clearly visible. The stars of the "Summer Triangle" (**Vega, Deneb** and **Altair**) in the constellations of **Lyra**, **Cygnus** and **Aquila**, respectively, are now readily seen in the east, and the small, distinctive constellation of **Delphinus** is well clear of the horizon.

Noctilucent clouds, photographed by Alan Tough from Nairn in Scotland, on May 31, 2020, at 00:28.

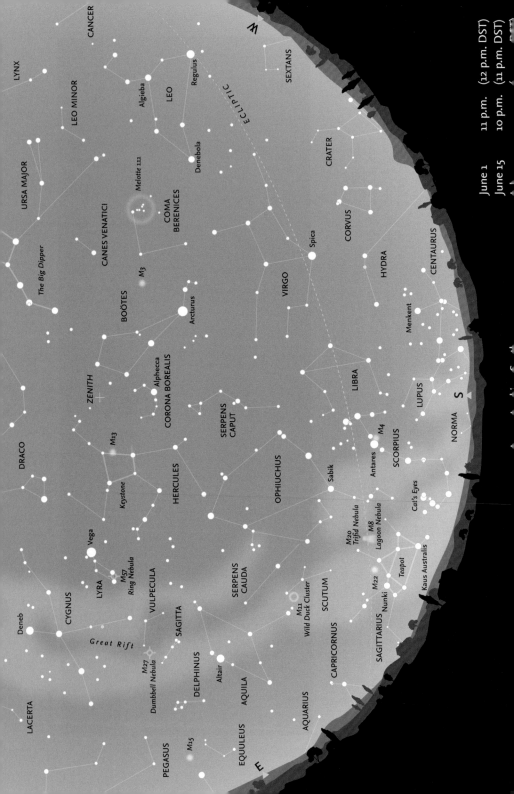

W

LYNX
CANCER
LEO MINOR
URSA MAJOR
The Big Dipper
CANES VENATICI
Algieba
LEO
Regulus
Melotte 111
COMA BERENICES
Denebola
SEXTANS
ECLIPTIC
CRATER
M3
BOÖTES
Arcturus
ZENITH
Alphecca
CORONA BOREALIS
Spica
VIRGO
CORVUS
HYDRA
CENTAURUS
Menkent
DRACO
SERPENS CAPUT
M13
Keystone
HERCULES
OPHIUCHUS
LIBRA
Sabik
LUPUS
NORMA
S
Vega
M57
Ring Nebula
LYRA
VULPECULA
SERPENS CAUDA
Cat's Eyes
M4
Antares
SCORPIUS
CYGNUS
Deneb
Great Rift
SAGITTA
M27
Dumbbell Nebula
M11
Wild Duck Cluster
SCUTUM
M20
Trifid Nebula
M8
Lagoon Nebula
Teapot
M22
Kaus Australis
LACERTA
DELPHINUS
Altair
AQUILA
CAPRICORNUS
SAGITTARIUS
Nunki
PEGASUS
M15
EQUULEUS
AQUARIUS
E

June 1   11 p.m.   (12 p.m. DST)
June 15   10 p.m.   (11 p.m. DST)

June – Looking South

The rather undistinguished constellation of **Libra** now lies almost due south. The red supergiant star **Antares** – the name means the "Rival of Mars" – in **Scorpius** is visible slightly to the east of the meridian, and even the "tail" or "sting" is visible with clear skies. Larger areas of **Centaurus** and **Lupus** are visible in the south, with **Sagittarius** to the east. Higher in the sky is the large constellation of **Ophiuchus** (the "Serpent Bearer"), lying between the two halves of **Serpens: Serpens Caput** ("Head of the Serpent") to the west and **Serpens Cauda** ("Tail of the Serpent") to the east. (Serpens is the only constellation to be divided into two distinct parts.) The ecliptic runs across Ophiuchus, and the Sun spends far more time in the constellation than it does in the "classical" zodiacal constellation of Scorpius, a small area of which lies between Libra and Ophiuchus.

Higher in the southern sky, the three constellations of **Boötes**, **Corona Borealis** and **Hercules** are now better placed for observation than at any other time of the year. This is an ideal time to observe the fine globular cluster of M13 in Hercules.

The Moon's phases for June 2024

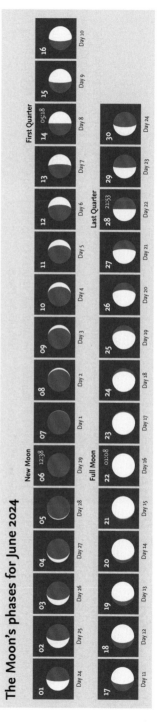

The constellations of Boötes and Corona Borealis are high in the sky in June. Arcturus has an orange tint and is the brightest star (mag. –0.05) in the northern celestial hemisphere.

June – Moon and Planets

The Moon

On June 1, the Moon passes extremely close and to the south of **Neptune**, but there is no occultation. The next day it is 2.4°N of **Mars**. On June 5, the day before New Moon, it passes 3.7°N of **Uranus** (mag. 5.8) and later that day 4.7°N of **Jupiter** (mag. -2.0). Later again, that same day, it is 4.7°N of **Mercury** (mag. -1.2). It is 9.9°N of **Aldebaran** on June 6 (at New Moon). By June 9 it is 1.7°S of **Pollux** and by June 12, 3.3°N of **Regulus** in **Leo**. On June 16 it is 1.2°N of **Spica**. By June 20, the Moon is 0.3°N of **Antares**. By June 27, the Moon is 0.1°N of **Saturn** and, the next day (June 28) it is 0.3°N of **Neptune** (mag. 7.9).

The planets

Mercury moves from the morning to the evening sky and is mag. -1.2 and 4.7°S of the Moon on June 5 (one day before New Moon). **Venus** is very close to the Sun in **Gemini**. Mars (mag. 1.0) moves from **Pisces** into **Aries**. Jupiter (mag. -2.0) is in **Taurus**, 4.7°S of the Moon on June 5 (one day before New Moon). **Saturn** (mag. 1.1) is moving slowly east in **Aquarius**. **Uranus** (mag. 5.8) is now in **Taurus** and **Neptune** (mag. 7.9) remains in **Pisces**.

The path of the Sun and the planets along the ecliptic in June.

Calendar for June

01	02:54	Neptune (mag.7.9) 0.0°N of the Moon
02	07:16	Moon at perigee = 368,102 km
02	23:37	Mars (mag.1.1) 2.4°S of the Moon
04	10:00 *	Jupiter (mag.-2.0) 0.1°N of Mercury (mag.-1.1)
05	00:37	Uranus (mag.5.8) 3.7°S of the Moon
05	14:25	Jupiter (mag.-2.0) 4.7°S of the Moon
05	18:28	Mercury (mag.-1.2) 4.7°S of the Moon
06	04:21	Aldebaran 9.9°S of the Moon
06	12:38	New Moon
09	08:00	Pollux 1.7°N of the Moon
12	03:41	Regulus 3.3°S of the Moon
14	05:18	First Quarter
14	13:25	Moon at apogee = 404,077 km
16	18:11	Spica 1.2°S of the Moon
20	11:11	Antares 0.3°S of the Moon
22	01:08	Full Moon
27	11:30	Moon at perigee = 369,286 km
27	15:00	Saturn (mag.1.1) 0.1°S of the Moon
		An occultation will be visible from parts of eastern Australia.
28	08:56	Neptune (mag.7.9) 0.3°S of the Moon
28	21:53	Last Quarter

These objects are close together for an extended period around this time.

Evening 9:30 p.m. (DST)

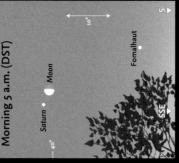

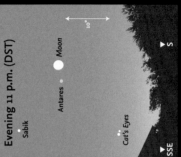

June 8 • *The Moon, Pollux and Castor form a right-angled triangle.*

Morning 5 a.m. (DST)

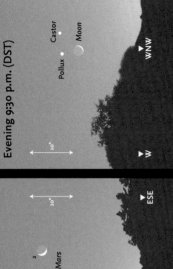

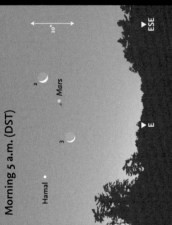

June 2–3 • *The waning crescent Moon passes Mars almost due east. Hamal is nearby.*

Evening 11 p.m. (DST)

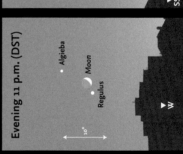

June 11 • *In the west, the Moon is between Regulus and Algieba.*

June 19 • *The Moon and Antares are close together. Sabik and the Cat's Eyes are farther east.*

Morning 5 a.m. (DST)

June 27 • *High in the southern sky, the Moon is close to Saturn. Fomalhaut is twenty degrees lower and farther south.*

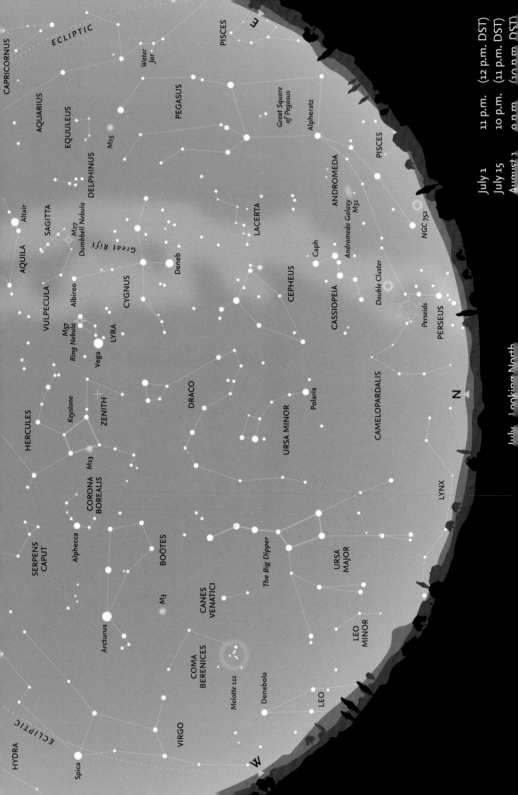

July · Looking North

July 1	11 p.m.	(12 p.m. DST)
July 15	10 p.m.	(11 p.m. DST)
August 1	9 p.m.	(10 p.m. DST)

July – Looking North

As in June, light nights and the chance of observing noctilucent clouds persist throughout July, but later in the month (and particularly after midnight) some of the major constellations begin to be more easily seen. **Capella**, the brightest star in **Auriga** together with the rest of the constellation, is hidden below the northern horizon. **Cassiopeia** is clearly visible in the northeast and **Perseus**, to its south, is beginning to climb clear of the horizon. The band of the Milky Way, from Perseus through Cassiopeia toward **Cygnus**, stretches up into the northeastern sky. If the sky is dark and clear, you may be able to make out the small, faint constellation of **Lacerta**, lying across the Milky Way between Cassiopeia and Cygnus. In the east, the stars of **Pegasus** are now well clear of the horizon, with the main line of stars forming **Andromeda** roughly parallel to the horizon low in the northeast. **Alpheratz** (α Andromedae) is actually the star at the northeastern corner of the **Great Square of Pegasus**. **Cepheus** and **Ursa Major** are on opposite sides of **Polaris** and **Ursa Minor**, in the east and west, respectively. The head of **Draco** is very close to the zenith (which is in **Hercules**) so the whole of this winding constellation is readily seen.

Meteors

July brings increasing meteor activity, mainly because there are several minor radiants active in the constellations of **Capricornus** and **Aquarius**. Because of their location, however, observing conditions are not particularly favorable for northern-hemisphere observers, although the first shower, the **α-Capricornids**, active from July 3 to August 15 (peaking July 30, with a tail to August 15), does often produce very bright fireballs. The maximum rate, however, is only about 5 per hour. The parent body is Comet 169P/NEAT. The most prominent shower is probably that of the **Southern δ-Aquariids**, which are active from around July 12 to August 23, with a peak on July 30, although even

Cygnus, sometimes known as the "Northern Cross," depicts a swan flying down the Milky Way towards Sagittarius. The brightest star, Deneb (α Cygni), represents the tail, and Albireo (β Cygni) marks the position of the head, and lies near the bottom of the image.

then the rate is unlikely to reach 25 meteors per hour. In this case, the parent body may be Comet 96P/Machholz. This year, both shower maxima occur when the Moon is a few days before New Moon so observing conditions are reasonably favorable. A chart showing the δ-Aquariids radiant is shown on page 30. The **Perseids** begin on July 17 and peak on August 12–13.

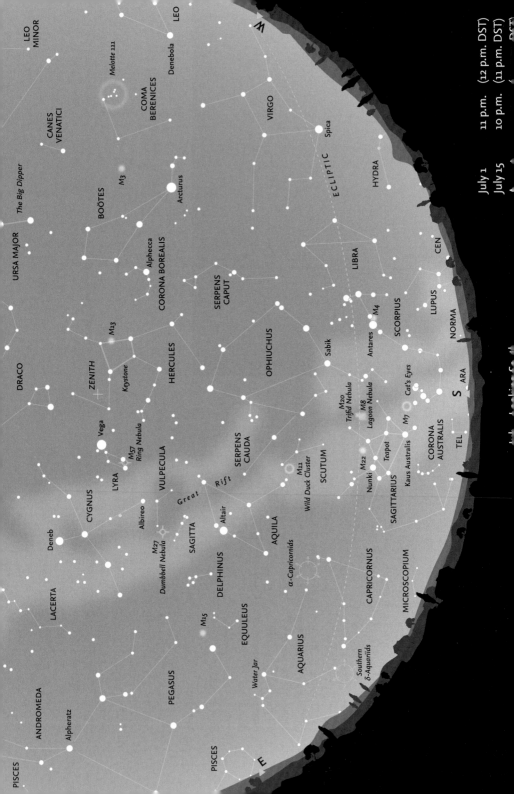

LEO
MINOR

LEO

Melotte 111

Denebola

COMA
BERENICES

CANES
VENATICI

The Big Dipper

VIRGO

Spica

ECLIPTIC

M3

BOÖTES

Arcturus

HYDRA

URSA MAJOR

Alphecca

CORONA BOREALIS

SERPENS
CAPUT

LIBRA

CEN

M13

DRACO

ZENITH

Keystone

HERCULES

OPHIUCHUS

Sabik

SCORPIUS

M4

Antares

LUPUS

NORMA

ARA

S

Vega

M57
Ring Nebula

LYRA

M20
Trifid Nebula

M8
Lagoon Nebula

Cat's Eyes

TEL

CYGNUS

Albireo

VULPECULA

SERPENS
CAUDA

Great Rift

SCUTUM

M11
Wild Duck Cluster

M7

Teapot

CORONA
AUSTRALIS

Deneb

M27
Dumbbell Nebula

SAGITTA

Altair

AQUILA

Nunki

M22

Kaus Australis

SAGITTARIUS

LACERTA

DELPHINUS

α-*Capricornids*

CAPRICORNUS

MICROSCOPIUM

PEGASUS

EQUULEUS

M15

AQUARIUS

Water Jar

ANDROMEDA

Alpheratz

PISCES

*Southern
δ-Aquariids*

PISCES

E

W

 N

July – Looking South

This is the best time of year to see **Scorpius**, with deep red **Antares** (α Scorpii), glowing just above the southern horizon. At around midnight, part of **Sagittarius**, with the distinctive asterism of the "Teapot," and the dense star clouds of the center of the Milky Way, are visible in the south, together with the small constellation of **Corona Australis**. The **Great Rift** – actually dust clouds that hide the more distant stars – runs down the Milky Way from **Cygnus** toward Sagittarius. Toward its northern end is the small constellation of **Sagitta** and the planetary nebula **M27** (the Dumbbell Nebula) in the otherwise insignificant constellation of **Vulpecula**. The sprawling constellation of **Ophiuchus** lies close to the meridian for a large part of the month, separating the two halves of the constellation of **Serpens**. The western half is called **Serpens Caput** (Head of the Serpent) and the eastern part **Serpens Cauda** (Tail of the Serpent). In the east, the bright **Summer Triangle**, consisting of **Vega** in **Lyra**, **Deneb** in Cygnus and **Altair** in **Aquila**, begins to dominate the southern sky, as it will throughout August and into September. The small constellation of Lyra, with Vega and a distinctive quadrilateral of stars to its east and south, lies not far from the zenith.

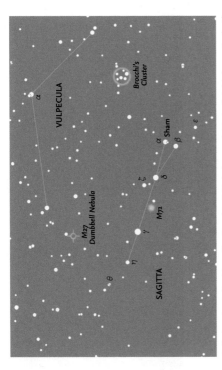

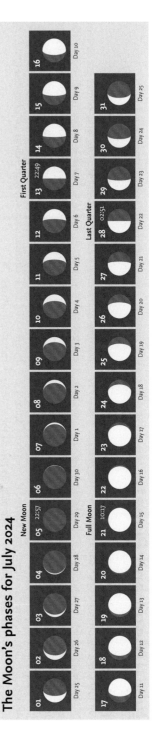

A finder chart for M27, the Dumbbell Nebula, a relatively bright (magnitude 8) planetary nebula – a shell of material ejected in the late stages of a star's lifetime – in the constellation of Vulpecula. All stars brighter than magnitude 7.5 are shown.

The Moon's phases for July 2024

July – Moon and Planets

The Moon

On July 1, the Moon is 4.1°N of **Mars** (mag.1.0). The next day it is 4.0°N of **Uranus** (mag.5.8). By July 3 it is 5.0°N of **Jupiter** (mag.-2.0) and, later, 9.9°N of **Aldebaran**. On July 6 the Moon is 3.9°N of **Venus** (mag.-3.9) and later 1.8°S of **Pollux** in **Gemini**. The next day the Moon is 3.2°N of **Mercury** (mag.-0.3). By July 9 it passes 3.0°N of **Regulus**, between it and **Algieba**. On July 14, one day after First Quarter, it is 0.9°N of **Spica**. On July 17, the Moon is 0.2°N of **Antares** in **Scorpius**. By July 24 the waning gibbous Moon is 0.4°N of **Saturn** (mag.0.9) and the next day 0.6°N of **Neptune** (mag.7.8). On July 29, one day after Last Quarter, the waning crescent Moon is 4.2°N of **Uranus** (mag.5.7). By July 30 it passes 10.1°N of **Aldebaran**, 5.0°N of **Mars** (mag.0.9) and 5.4°N of **Jupiter** (mag.-2.1).

The planets

Mercury is moving eastwards past the Sun, and comes to eastern elongation on July 22 at mag.0.3. **Venus** is bright (mag.-3.9) in the evening sky. **Mars** (mag.1.0 to 0.9) is in **Aries**, moving into **Taurus** by the end of the month. **Jupiter** (mag.-2.0 to -2.1) is moving slowly eastwards in **Taurus**. **Saturn** (mag.1.0 to 0.9) is in **Aquarius** and begins retrograde motion on July 4. **Uranus** (mag.5.7) is in **Taurus** and **Neptune** (mag.7.9) remains in **Pisces**.

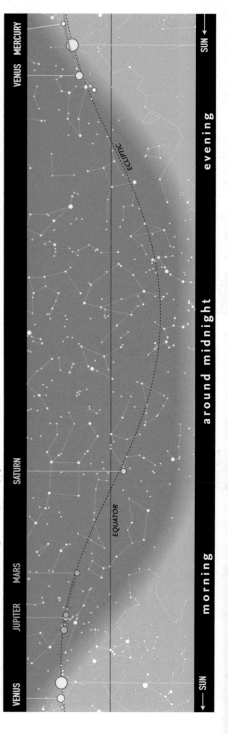

The path of the Sun and the planets along the ecliptic in July.

Calendar for July

Morning 5 a.m. (DST)

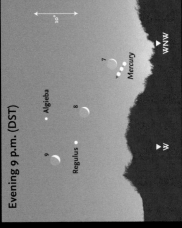

July 1–3 • Shortly before sunrise, the waning crescent Moon passes Mars, the Pleiades, Jupiter and Aldebaran and Elnath.

Evening 9 p.m. (DST)

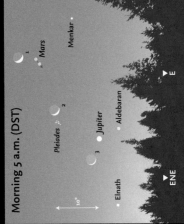

July 13 • The First Quarter Moon is close to Spica, in the southwestern sky.

Evening 9 p.m. (DST)

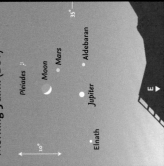

July 7–9 • On July 7, the Moon is close to Mercury. Two days later it has passed between Regulus and Algieba.

After midnight 2 a.m. (DST)

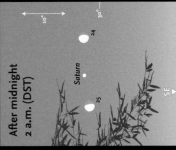

July 24–25 • The Moon passes Saturn (mag. 0.9) almost due southeast.

Morning 5 a.m. (DST)

July 30 • A nice gathering of the Moon, Jupiter, Mars, Aldebaran, the Pleiades and Elnath, shortly before sunrise.

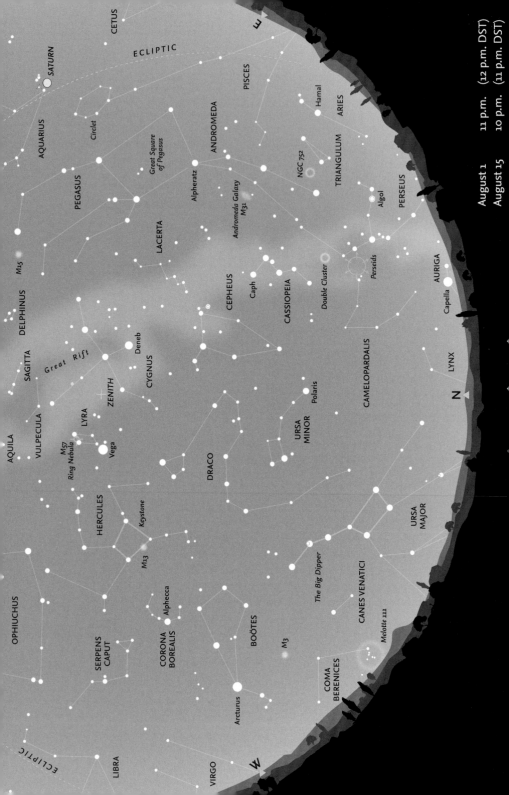

CETUS

E

ECLIPTIC

SATURN

PISCES

AQUARIUS

Circlet

PEGASUS

Great Square
of Pegasus

ANDROMEDA

Alpheratz

LACERTA

Andromeda Galaxy
M31

Hamal

ARIES

TRIANGULUM

NGC 752

PERSEUS

Algol

Perseids

AURIGA

Capella

M15

DELPHINUS

SAGITTA

Great Rift

Deneb

CYGNUS

ZENITH

CEPHEUS

Caph

CASSIOPEIA

Double Cluster

CAMELOPARDALIS

LYNX

N

AQUILA

VULPECULA

LYRA

M57
Ring Nebula

Vega

DRACO

Polaris

URSA
MINOR

HERCULES

Keystone

M13

URSA
MAJOR

OPHIUCHUS

SERPENS
CAPUT

CORONA
BOREALIS

Alphecca

BOÖTES

M3

The Big Dipper

CANES VENATICI

Melotte 111

COMA
BERENICES

Arcturus

LIBRA

VIRGO

W

ECLIPTIC

August 1   11 p.m. (12 p.m. DST)
August 15   10 p.m. (11 p.m. DST)

August – Looking North

A brilliant Perseid fireball, streaking alongside the Great Rift in the Milky Way, photographed in 2012 by Jens Hackmann from near Weikersheim in Germany. Four additional, fainter Perseids are also visible in the image.

Ursa Major is now the "right way up" in the northwest, although some of the fainter stars in the south of the constellation are difficult to see. Beyond it, *Boötes* stands almost vertically in the west, but pale orange *Arcturus* is sinking toward the horizon. Higher in the sky, both *Corona Borealis* and *Hercules* are clearly visible.

In the northeast, *Capella* becomes visible later in the night, but most of *Auriga* still remains below the horizon. Higher in the sky, *Perseus* is gradually coming into full view and, later in the night and later in the month, the beautiful *Pleiades* cluster rises above the northeastern horizon. Between Perseus and *Polaris* lies the faint and unremarkable constellation of *Camelopardalis*.

Higher still, both *Cassiopeia* and *Cepheus* are well placed for observation, despite the fact that Cassiopeia is completely immersed in the band of the Milky Way, as is the "base" of Cepheus. *Pegasus* and *Andromeda* are now well above the eastern horizon and, below them, the constellation of *Pisces* is climbing into view. Two of the stars in the "Summer Triangle," *Deneb* and *Vega*, are close to the zenith high overhead.

Meteors

August is the month when one of the best meteor showers of the year occurs: the *Perseids*. This is a long shower, generally beginning about July 17 and continuing until around August 24, with a maximum on August 12–13, when the rate may reach as high as 100 meteors per hour (and on rare occasions, even higher). In 2024, maximum is around First Quarter, so conditions are reasonably favorable. The Perseids are debris from Comet 109P/Swift-Tuttle (the Great Comet of 1862). Perseid meteors are fast and many of the brighter ones leave persistent trains. Some bright fireballs also occur during the shower.

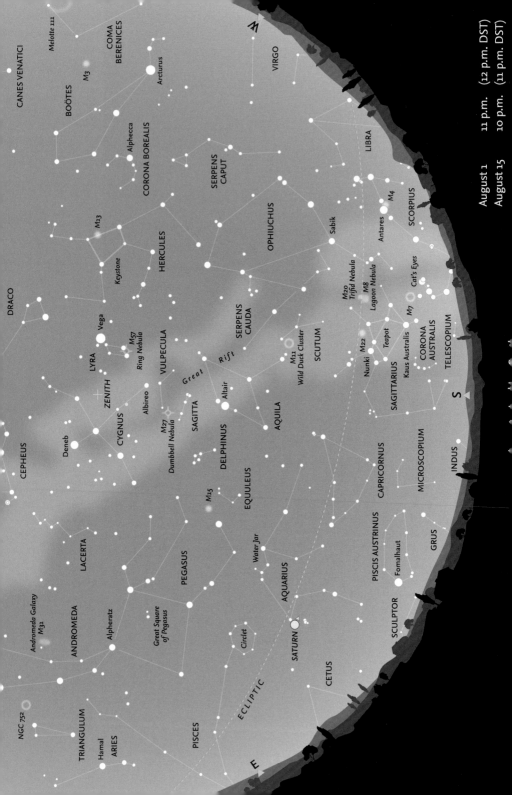

CANES VENATICI

Melotte 111

COMA
BERENICES

M3

BOÖTES

Arcturus

VIRGO

Alphecca

CORONA BOREALIS

LIBRA

SERPENS
CAPUT

M4

M13

OPHIUCHUS

SCORPIUS

DRACO

Antares

HERCULES

Sabik

Keystone

M20
Trifid Nebula

Cat's Eyes

Vega

M57
Ring Nebula

VULPECULA

M8
Lagoon Nebula

M7

LYRA

SERPENS
CAUDA

ZENITH

Great Rift

Albireo

M11
Wild Duck Cluster

SCUTUM

CYGNUS

M27
Dumbbell Nebula

SAGITTA

Altair

Teapot

S

Deneb

AQUILA

M22

Nunki

SAGITTARIUS

CORONA
AUSTRALIS

CEPHEUS

DELPHINUS

Kaus Australis

TELESCOPIUM

EQUULEUS

MICROSCOPIUM

LACERTA

M15

INDUS

CAPRICORNUS

Andromeda Galaxy
M31

PEGASUS

Water Jar

GRUS

ANDROMEDA

Alpheratz

PISCIS AUSTRINUS

Great Square
of Pegasus

AQUARIUS

Fomalhaut

NGC 752

Circlet

SCULPTOR

TRIANGULUM

SATURN

Hamal
ARIES

PISCES

ECLIPTIC

CETUS

E

August 1   11 p.m.   (12 p.m. DST)
August 15   10 p.m.   (11 p.m. DST)

August – Looking South

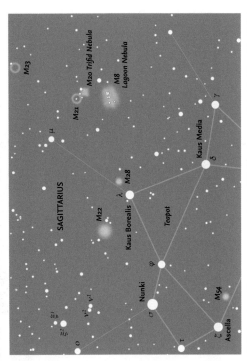

The whole stretch of the summer Milky Way stretches across the sky in the south, from **Cygnus**, high in the sky near the zenith, past **Aquila**, with bright **Altair** (α Aquilae), to part of the constellation of **Sagittarius** close to the horizon, where the pattern of stars known as the "Teapot" is visible. This area contains many nebulae and both open and globular clusters. Between **Albireo** (β Cygni) and Altair lie the two small constellations of **Vulpecula** and **Sagitta**, with the latter easier to distinguish (because of its shape) from the clouds of the Milky Way. Between Sagitta and **Pegasus** to the east lie the highly distinctive five stars that form the tiny constellation of **Delphinus** (again, one of the few constellations that actually bear some resemblance to the creatures after which they are named). Below Aquila, mainly in the star clouds of the Milky Way, lies **Scutum**, most famous for the bright open cluster, **M11** or the "Wild Duck Cluster", readily visible in binoculars. To the southeast of Aquila lie the two zodiacal constellations of **Capricornus** and **Aquarius** and, farther south, the constellation of **Piscis Austrinus** with the brilliant star **Fomalhaut**.

The Moon's phases for August 2024

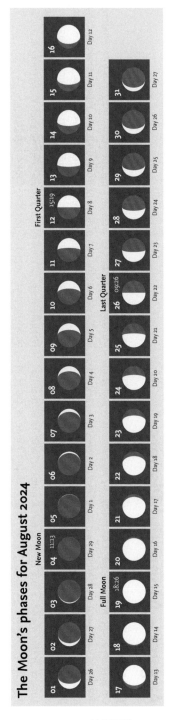

A finder chart for the gaseous nebulae M8 (the Lagoon Nebula), M20 (the Trifid Nebula) and the globular cluster M22, all in Sagittarius. Clusters M21, M23 (open) and M28 (globular) are faint. The chart shows all stars brighter than magnitude 7.5.

August – Moon and Planets

The Moon

On August 2, the Moon is 1.8°S of *Pollux*. By August 5 it is 2.9°N of *Regulus* in *Leo*. Later the same day it is 1.7°N of *Venus* (mag. -3.8). On August 6 the Moon is 7.5°N of *Mercury* and close to *Venus*. By August 10 the waxing crescent Moon is 0.7°N of *Spica*. On August 14, the Moon occults *Antares*, but this is visible only over the Pacific Ocean. By August 21, the Moon is 0.5°N of *Saturn* (mag. 0.7) and later that day is 0.7°N of *Neptune* (at mag. 7.8). By August 26 the Moon is 4.4°N of *Uranus* (mag. 5.7) and, later, 10.3°N of *Aldebaran*. On August 28, the Moon passes 5.3°N of *Mars* and on August 30, 1.7°S of *Pollux*.

The Planets

Mercury rapidly moves west and declines from mag. 0.9 to 2.2 before brightening at the end of the month to meg 0.8. *Venus* is close by early in the month. *Mars* (mag. 0.8) is moving east in *Taurus*. *Jupiter* is also moving slowly east in *Taurus*. *Saturn* (mag. 0.7) continues retrograde motion throughout August. *Uranus* (mag. 5.7) is still in *Taurus* and *Neptune* (mag. 7.8) remains in *Pisces*. Minor planet (7) Iris (mag. 8.1) is at opposition on August 6 (see the maps on page 27).

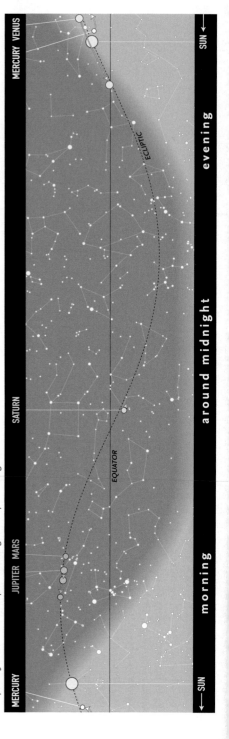

The path of the Sun and the planets along the ecliptic in August.

Calendar for August

02	23:35	Pollux 1.8°N of the Moon
04	11:13	New Moon
05	19:34	Regulus 2.9°S of the Moon
05	22:03	Venus (mag. -3.8) 1.7°S of the Moon
06	00:02	Mercury (mag. 1.6) 7.5°S of the Moon
06	15:00 *	Venus (mag. -3.8) 5.9°N of Mercury (mag. 1.6)
06	19:40	Minor planet (7) Iris at opposition (mag. 8.1)
09	01:31	Moon at apogee = 405,297 km
09	10:17	Spica 0.7°S of the Moon
10		Perseid meteor shower maximum
12–13		
12	15:19	First Quarter
14	05:17	Antares 0.0°N of the Moon
19	18:26	Full Moon
21	03:02	Saturn (mag. 0.7) 0.5°N of the Moon
21	05:02	Moon at perigee = 360,196 km
21	22:21	Neptune (mag. 7.8) 0.7°S of the Moon
26	00:01	Uranus (mag. 5.7) 4.4°S of the Moon
26	09:26	Last Quarter
26	23:29	Aldebaran 10.3°S of the Moon
27	12:43	Jupiter (mag. -2.3) 5.7°S of the Moon
28–Sep.05		α-Aurigid meteor shower
28	00:22	Mars (mag. 0.8) 5.3°S of the Moon
30	05:25	Pollux 1.7°N of the Moon

* These objects are close together for an extended period around this time.

Evening 9 p.m. (DST)

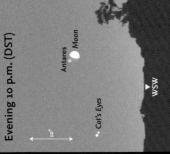

August 9–10 • The waxing crescent Moon passes Spica, in the southwestern sky.

Evening 10 p.m. (DST)

August 13 • The Moon and Antares are close together. The Cat's Eyes are farther south.

Evening 10 p.m. (DST)

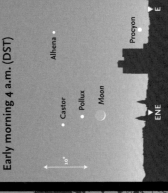

August 20 • The Moon is close to Saturn (mag 0.7), in the east-southeast.

Morning 5 a.m. (DST)

August 26–27 • Just after Last Quarter, the Moon passes the Pleiades, Aldebaran, Jupiter, Mars and Elnath. Capella, Menkalinan (β Aur) and Bellatrix (γ Ori) are all nearby.

Early morning 4 a.m. (DST)

August 30 • The narrow crescent Moon is near Pollux and Castor. Alhena (γ Gem) and Procyon are farther east.

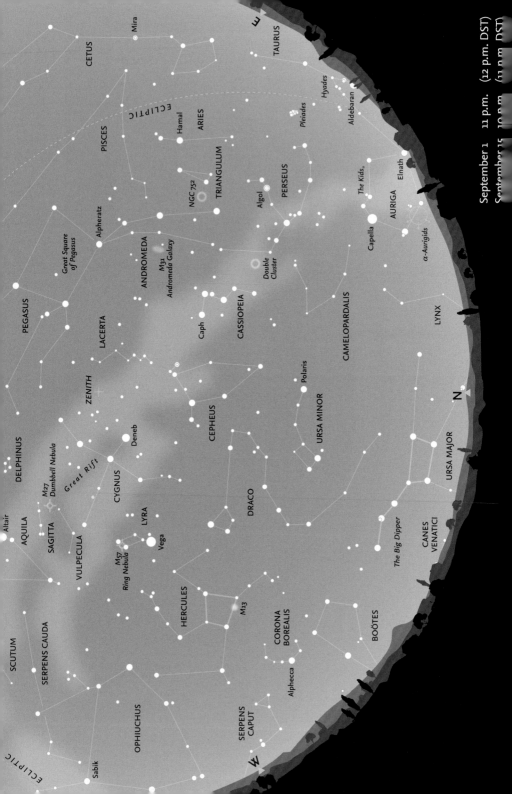

September – Looking North

The (northern) autumnal equinox occurs on September 22, when the Sun moves south of the equator in Virgo.

Ursa Major is now low in the north, and to the northwest **Arcturus** and much of **Boötes** have sunk below the horizon. In the northeast, **Auriga** is beginning to climb higher in the sky. Later in the month, **Taurus**, with orange **Aldebaran** (α Tauri), and even **Gemini**, with **Castor** and **Pollux**, become visible in the east and northeast. Due east, **Andromeda** is now easily visible, with the small constellations of **Triangulum** and **Aries** (the latter a zodiacal constellation) directly below it. Practically the whole of the northern Milky Way is visible, arching across the sky, both in the north and in the south. It is not particularly clear in Auriga, or even **Perseus**, but in **Cassiopeia** and on toward **Cygnus** the clouds of stars become easier to see. The **Double Cluster** in Perseus is well placed for observation. **Cepheus** is "upside-down" near the zenith, and the head of **Draco** and **Hercules** beyond it are well placed for observation.

Meteors

After the major Perseid shower in August, there is very little shower activity in September. One minor, but very extended, shower, known as the **α-Aurigids,** tends to have two peaks of activity. The principal peak occurs on September 1. In 2024, the Moon is two days before New Moon, so conditions are very favorable. At maximum, however, the hourly rate hardly reaches 10 meteors per hour, although the meteors are bright and relatively easy to photograph. Activity from this shower may even extend into October. The **Southern Taurid** shower begins this month (on September 10) and, although rates are low, often produces very bright fireballs. This is a very long shower, lasting until about November 20. As a slight compensation for the lack of shower activity, however, in September the number of sporadic meteors reaches its highest rate at any time during the year.

The twin open clusters, known as the Double Cluster, in Perseus (more formally called h and χ Persei) are close to the main portion of the Milky Way.

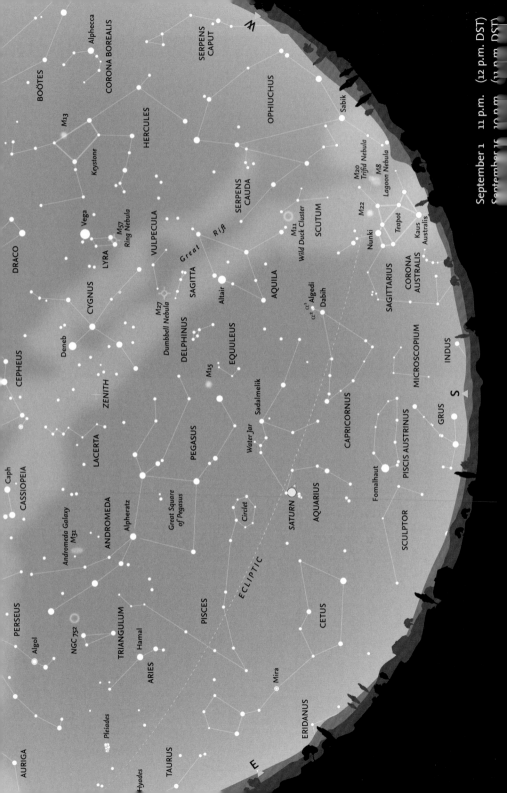

September – Looking South

The "Summer Triangle" is now high in the southwest, with the Great Square of **Pegasus** high in the southeast. Below Pegasus are the two zodiacal constellations of **Capricornus** and **Aquarius**. In what is otherwise an unremarkable constellation, **Algedi** (α Capricorni) is actually a visual binary, with the two stars (α¹ Cap and α² Cap) readily seen with the naked eye. **Dabih** (β Capricorni), just to the south, is also a double star, and the components are relatively easy to separate with binoculars. In Aquarius, just to the east of **Sadalmelik** (α Aquarii) there is a small asterism consisting of four stars, resembling a tiny letter "Y", known as the "Water Jar." Below Aquarius is a sparsely populated area of the sky with just one bright star in the constellation of **Piscis Austrinus**. In classical illustrations, water is shown flowing from the "Water Jar" towards bright **Fomalhaut** (α Piscis Austrini).

Another zodiacal constellation, **Pisces**, is now clearly visible to the east of Aquarius. Although faint, there is a distinctive asterism of stars, known as the "Circlet," south of the Great Square and another line of faint stars to the east of Pegasus. Still farther down towards the horizon is the constellation of **Cetus**, with the famous variable star **Mira** (ο Ceti) at its

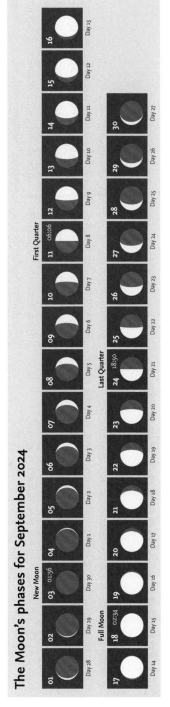

The constellations of Pisces and Aries. The 'Circlet' of Pisces is bottom right and the three main stars of Aries, top left. In 2024, Jupiter is in Aries until the end of April.

center. When Mira is at maximum brightness (around mag. 3.5) it is clearly visible to the naked eye, but it disappears as it fades towards minimum (about mag. 9.5 or less). There is a finder chart for Mira on page 97.

The Moon's phases for September 2024

New Moon

01	02	03 01:56
Day 28	Day 29	Day 30

Full Moon

04	05	06	07	08	09	10	11 06:06	12	13	14	15	16
Day 1	Day 2	Day 3	Day 4	Day 5	Day 6	Day 7	Day 8	Day 9	Day 10	Day 11	Day 12	Day 13

First Quarter

17	18 02:34	19	20	21	22	23	24 18:50	25	26	27	28	29	30
Day 14	Day 15	Day 16	Day 17	Day 18	Day 19	Day 20	Day 21	Day 22	Day 23	Day 24	Day 25	Day 26	Day 27

Last Quarter

September – Moon and Planets

The Moon

On September 5, the waxing crescent Moon (two days after New) is 1.2°S of *Venus* (mag.-3.8). By September 17, one day before Full, the Moon is 0.3°N of *Saturn*. At Full Moon it is 0.7°N of *Neptune* (mag.7.8) On September 22, it is 4.5°N of *Uranus* (mag.5.6) and the next day, September 23 just before Last Quarter, it is 5.7°N of brilliant *Jupiter* (mag.-2.4). On September 25 it is 4.9°N of *Mars* (mag.0.6).

The Planets

Mercury (mag. 0.9 to -1.7) reaches greatest western elongation on September 5 (see diagram on page 22). *Venus* (mag. -3.8) moves from *Virgo* into *Libra*, well east of the Sun in the evening sky. *Mars* (mag. 0.6) moves from *Taurus* into *Gemini*. *Jupiter* is moving slowly east in *Taurus*. *Saturn* is still retrograding in *Aquarius* and comes to opposition on September 8. *Uranus* (mag. 5.6) begins retrograde motion on September 14 in *Aquarius*. *Neptune* (mag. 7.8) is in *Pisces* and is at opposition on September 21 (see the maps on page 25).

The path of the Sun and the planets along the ecliptic in September.

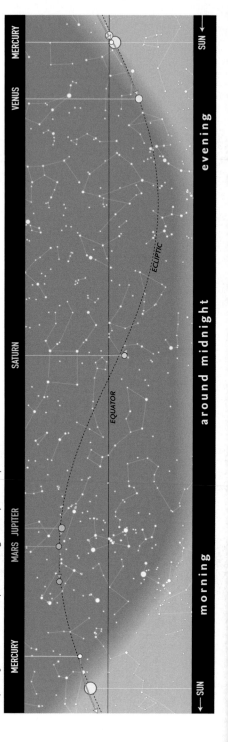

Calendar for September

01		α-Aurigid meteor shower maximum
01	09:16	Mercury (mag. 0.5) 5.0°S of the Moon
03	01:56	New Moon
05	02:30	Mercury at greatest elongation (18.1°W, mag. -0.3)
05	10:16	Venus (mag. -3.8) 1.2°N of the Moon
05	14:54	Moon at apogee = 406,211 km
08	04:35	Saturn at opposition (mag. 0.6)
10–Nov.20		Southern Taurid meteor shower
11	06:06	First Quarter
17	10:22	Saturn (mag. 0.6) 0.3°S of the Moon
18	02:34	Full Moon
18	02:45	Partial lunar eclipse
18	07:35	Neptune (mag. 7.8) 0.7°S of the Moon
18	13:22	Moon at perigee = 357,286 km
21	00:17	Neptune at opposition (mag. 7.8)
22	07:14	Uranus (mag. 5.6) 4.5°S of the Moon
22	12:44	September equinox
23	23:21	Jupiter (mag. -2.4) 5.7°S of the Moon
24	18:50	Last Quarter
25	11:49	Mars (mag. 0.5) 4.9°S of the Moon

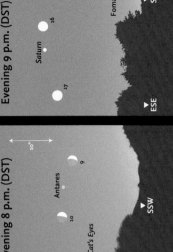

Evening 8 p.m. (DST)

September 5 • After sunset, the Moon, Venus and Spica are near the western horizon.

Evening 8 p.m. (DST)

September 9–10 • The Moon passes Antares. The Cat's Eyes are farther south.

Evening 9 p.m. (DST)

September 16–17 • The Moon passes Saturn. Fomalhaut is closer to the horizon and farther south.

Morning 6 a.m. (DST)

September 22 • High in the southwest, the Moon is close to the Pleiades. Aldebaran is closeby.

Early morning 4 a.m. (DST)

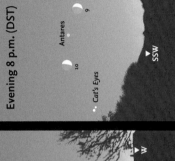
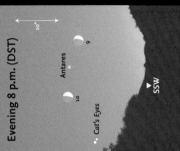

September 23–25 • The Moon passes Aldebaran, Elnath and Mars, high in the east.

Morning 6 a.m. (DST)

September 26 • The Moon lines up with Pollux and Castor. Mars is farther south.

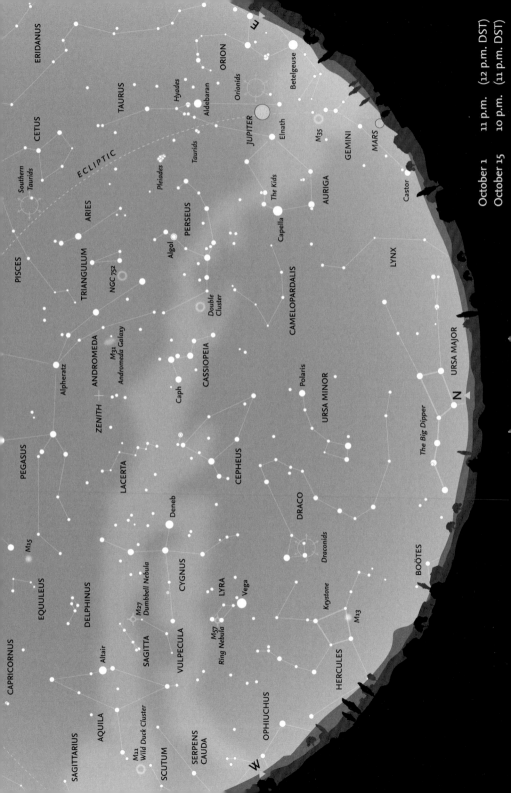

October – Looking North

Ursa Major is grazing the horizon in the north, while high overhead are the constellations of *Cepheus*, *Cassiopeia* and *Perseus*, with the Milky Way between Cepheus and Cassiopeia below the zenith. *Auriga* is now clearly visible in the east, as is *Taurus* with the *Pleiades*, *Hyades* and orange *Aldebaran*. Also in the east, *Orion* and *Gemini* are starting to rise clear of the horizon.

The constellations of *Boötes* and *Corona Borealis* are now lost to view in the northwest, and *Hercules* is also descending toward the western horizon. The three stars of the "Summer Triangle" are still clearly visible, although *Aquila* and *Altair* are beginning to approach the horizon in the west. Toward the end of the month (October 27) Summer Time ends in Europe. In North America, the end of DST comes next month, in November.

Meteors

The *Orionids* are the major, fairly reliable meteor shower active in October. Like the May *η-Aquariid* shower, the Orionids are associated with Comet 1P/Halley. During this second pass through the stream of particles from the comet, slightly fewer meteors are seen than in May, but conditions are more favorable for northern observers. In both showers the meteors are very fast, and many leave persistent trains. Although the Orionid maximum is quoted as October 21–22, in fact there is a very broad maximum, lasting about a week roughly centered on that date, with hourly rates around 25. Occasionally, rates are higher (50–70 per hour). In 2024, the Moon is waning gibbous, so conditions are not particularly favorable. The faint shower of the *Southern Taurids* (often with bright fireballs) peaks on October 10–11. The Southern Taurid maximum occurs when the Moon is at around First Quarter, so conditions are more favorable than those for the Orionids. Toward the

The constellation of Perseus is not only the location of the radiant of the Perseid meteor shower in August, but is also well known for the pair of open clusters, called the Double Cluster (near the top edge of the image), close to the border with Cassiopeia, and also for Algol (β Persei), the famous variable star (just below the center of the photograph).

end of the month (around October 20), another shower (the **Northern Taurids**) begins to show activity, which peaks early in November. The parent comet for both Taurid showers is Comet 2P/Encke. The meteors in both Taurid streams are relatively slow and bright.

A minor shower, the **Draconids**, begins on October 6 and peaks on October 8–9.

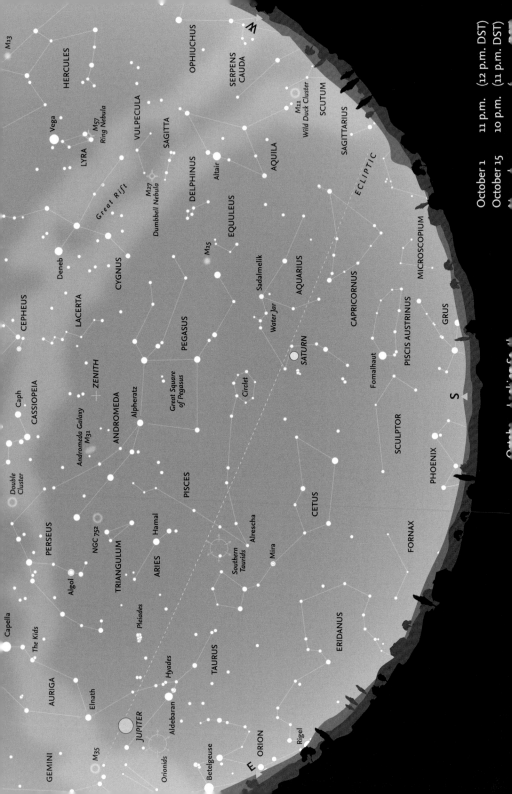

October – Looking South

The Great Square of **Pegasus** dominates the southern sky, framed by the two chains of stars that form the constellation of **Pisces**, together with **Alrescha** (α Piscium) at the point where the two lines of stars join. Also clearly visible is the constellation of **Cetus**, below Pegasus and Pisces. Although **Capricornus** is now lower, **Aquarius** to its east is well placed in the south, with solitary **Fomalhaut** and the constellation of **Piscis Austrinus** beneath it. The inconspicuous constellation of **Sculptor** appears in the south, as well as parts of **Grus**, **Phoenix** and, farther east, the northernmost stars of **Eridanus**.

The main band of the Milky Way and the Great Rift runs down from **Cygnus**, through **Vulpecula**, **Sagitta** and **Aquila** toward the western horizon. **Delphinus** and the tiny, unremarkable constellation of **Equuleus** lie between the band of the Milky Way and Pegasus. **Andromeda** is clearly visible high in the sky to the southeast, with the small constellation of **Triangulum** and the zodiacal constellation of **Aries** below it. **Perseus** is high in the east, and by now the **Pleiades** and **Taurus** are well clear of the horizon. Later in the night, and later in the month, **Orion** rises in the east, a sign that the autumn season has arrived and of the steady approach of winter.

The Moon's phases for October 2024

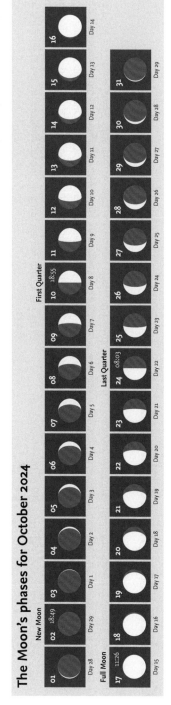

The constellation of Aquarius is one of the constellations that is visible in late summer and early autumn. The four stars forming the "Y"-shape of the "Water Jar" may be seen to the east of Sadalmelik (α Aquarii), the brightest star (top center).

October – Moon and Planets

The Moon

On October 2 there is an annular solar eclipse, visible over the Pacific Ocean (see pages 20 and 21). The next day the Moon is 1.8°S of **Mercury** (mag.-1.6). On October 5 the Moon is 3.0°S of **Venus** (mag.-3.9). By October 14 it is 0.1°N of **Saturn** (mag.0.7). On October 15 the Moon is 0.6°N of **Neptune** (mag.7.8) and on October 19 it is 4.5°N of **Uranus** (mag.5.6). On October 23 the Moon is 3.9°N of **Mars** (mag.0.2).

The planets

Mercury (mag.-1.6) moves towards the Sun. **Venus** (mag.-3.9 to -4.0) is moving eastwards in **Virgo**. **Mars** (changing rapidly from mag.0.5 to mag.0.1) is in **Gemini**, moving towards **Cancer**. **Jupiter** (mag.-2.5 to -2.7) begins retrograde motion on October 9 and is in **Taurus**. **Saturn** (mag. 0.7 to 0.8) is still retrograding slowly in **Aquarius**. **Uranus** (mag.5.6) is also retrograding in **Taurus**. **Neptune** (mag.7.8) remains in **Pisces**.

The path of the Sun and the planets along the ecliptic in October.

Calendar for October

02	18:46	Annular solar eclipse (Pacific)
02	18:49	New Moon
02	19:39	Moon at apogee = 406,516 km (most distant of the year)
02–Nov.07		Orionid meteor shower
03	00:02	Mercury (mag.-1.6) 1.8°N of the Moon
05	20:26	Venus (mag.-3.9) 3.0°N of the Moon
06–10		Draconid meteor shower
08–09		Draconid meteor shower maximum
10–11		Southern Taurid meteor shower maximum
10	18:55	First Quarter
14	18:13	Saturn (mag.0.7) 0.1°S of the Moon
15	17:32	Neptune (mag.7.8) 0.6°S of the Moon
17	11:26	Full Moon
17	22:50	Moon at perigee = 357,175 km
19	15:52	Uranus (mag.5.6) 4.5°S of the Moon
20–Dec.10		Northern Taurid meteor shower
21–22		Orionid meteor shower maximum
21	08:04	Jupiter (mag.-2.6) 5.8°S of the Moon
23	19:55	Mars (mag.0.2) 3.9°S of the Moon
24	08:03	Last Quarter
27		British Summer Time ends (UK reverts to GMT)

Evening 10 p.m. (DST)

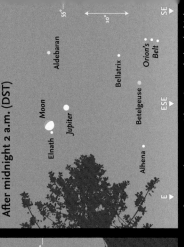

October 13–14 • The Moon passes Saturn. Fomalhaut and Diphda (β Cet) are closer to the horizon.

After midnight 2 a.m. (DST)

October 21 • The Moon with Jupiter and Elnath, high in the east-southeast. Aldebaran, Alhena and the brightest stars of Orion are also in the area.

After midnight 2 a.m. (DST)

October 23 • The Moon is almost at Last Quarter when it is near Pollux and Castor. Mars is about seven degrees lower. Alhena and Procyon are nearby...

Morning 5 a.m. (DST)

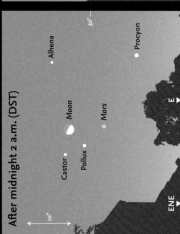

October 26 • The Moon with Regulus and Algieba, at an altitude of thirty-five degrees. Denebola is twenty degrees closer to the horizon...

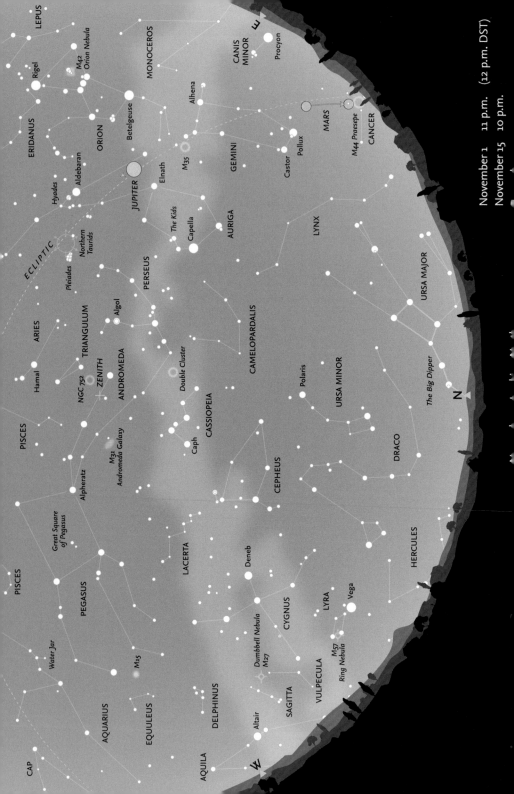

November – Looking North

The constellation of Auriga, with brilliant Capella (mag. 0.08), which, although appearing as a single star, is actually a quadruple system, consisting of a pair of yellow giant stars, gravitationally bound to a more distant pair of red dwarfs.

Most of **Aquila** has now disappeared below the horizon, but two of the stars of the "Summer Triangle," **Vega** in **Lyra** and **Deneb** in **Cygnus**, are still clearly visible in the west. The head of **Draco** is now low in the northwest and only a small portion of **Hercules** remains above the horizon. The southernmost stars of **Ursa Major** are now coming into view. The Milky Way arches overhead, with the denser star clouds in the west and the less heavily populated region through **Auriga** and **Monoceros** in the east. High overhead, **Cassiopeia** is near the zenith and **Cepheus** has swung round to the northwest, while Auriga is now high in the northeast. **Gemini**, with **Castor** and **Pollux**, is well clear of the eastern horizon, and even **Procyon** (α Canis Minoris) is just climbing into view almost due east.

Meteors

The **Northern Taurid** shower, which began in mid-October, reaches maximum – although with just a low rate of about five meteors per hour – on November 12–13. The Moon is waxing gibbous, so conditions are not very favorable. The shower gradually trails off, ending around December 10. There is an apparent 7-year periodicity in fireball activity, but 2024 is unlikely to be a peak year. Far more striking, however, are the **Leonids**, which have a relatively short period of activity (November 6–30), with maximum on November 18. This shower is associated with Comet 55P/Tempel-Tuttle and has shown extraordinary activity on various occasions with many thousands of meteors per hour. High rates were seen in 1999, 2001 and 2002 (reaching about 3,000 meteors per hour) but have fallen dramatically since then. The rate in 2024 is likely to be about 10 per hour. These meteors are the fastest shower meteors recorded (about 70 km per second) and often leave persistent trains. The shower is very rich in faint meteors. In 2024, maximum is when the Moon is waning gibbous, so conditions are not particularly favorable

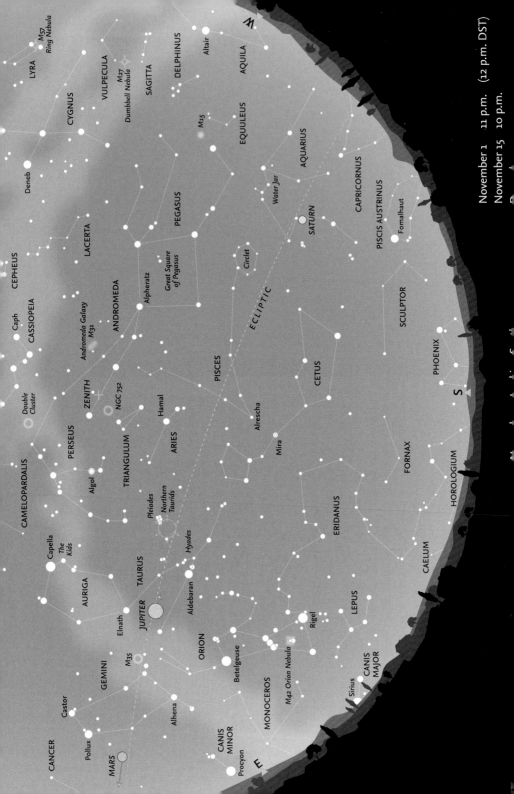

November 1   11 p.m.   (12 p.m. DST)
November 15   10 p.m.

November – Looking South

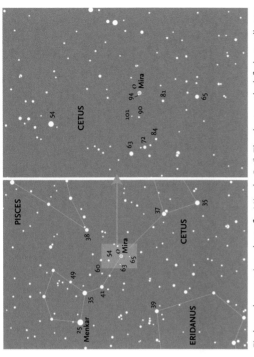

Orion has now risen above the eastern horizon, and much of the long, straggling constellation of **Eridanus** (which begins near **Rigel**) is visible to the west of Orion as is the small constellation of **Fornax**. Parts of **Horologium** and **Phoenix** are peeping above the southern horizon. Higher in the sky, **Taurus**, with the **Pleiades** cluster, and orange **Aldebaran** are now easy to observe. To their west, both **Pisces** and **Cetus** are close to the meridian. The famous long-period variable star, **Mira** (ο Ceti), with a typical range of magnitude 3.4 to 9.8, is favorably placed for observation. In the southwest, **Capricornus** has slipped below the horizon, but **Aquarius** remains visible. Even farther west, **Altair** may be seen early in the night, but most of **Aquila** has already disappeared from view. **Delphinus**, together with **Sagitta** and **Vulpecula** in the Milky Way, will soon vanish for another year. Both **Pegasus** and **Andromeda** are easy to see, and one of the lines of stars that make up Andromeda finishes close to the zenith, which is also close to one of the outlying stars of **Perseus**, high in the east.

The Moon's phases for November 2024

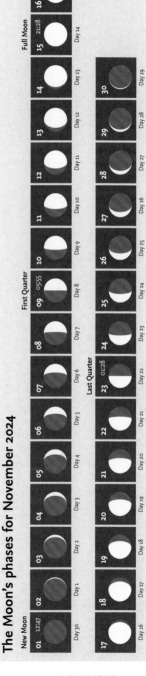

Finder and comparison charts for Mira (ο Ceti). The chart on the left shows all stars brighter than magnitude 6.5. The chart on the right shows stars down to magnitude 10.0. The comparison star magnitudes are shown without the decimal point.

November – Moon and Planets

The Moon

On November 3, the Moon is 2.1°S of *Mercury* (mag. -0.3). On November 4 |the Moon is 0.1°S of *Antares* and the next day (at 00:15) is 3.1°S of *Venus* (mag. -4.0). By November 12 the Moon is 0.6°N of *Neptune* (mag. 7.8). On November 16 the Moon is 4.4°N of *Uranus* (mag. 5.6). On November 17 the Moon is 10.3°N of *Aldebaran*, later that day it is 5.5°N of *Jupiter* (mag. -2.8). By November 20, the Moon is 1.9°S of *Pollux*, later that day it is 2.4°N of *Mars* in *Cancer*. By November 22, the Moon passes 2.7°N of *Regulus* between it and *Algieba*. On November 27 the Moon is 0.4°N of *Spica* in *Virgo*.

The Planets

Mercury reaches eastern elongation on November 16. *Venus* is very bright (mag. -4.0 to -4.2) and moving eastwards away from the Sun. *Mars* (mag. 0.1 to -0.4) moves eastwards in *Cancer*. *Jupiter* (mag. -2.7 to -2.8) is retrograding in *Taurus*. *Saturn* (mag. 0.8 to 1.0) is in *Aquarius*. *Uranus* (mag. 5,6) is in *Taurus* and comes to opposition on November 17 (see the maps on page 25). *Neptune* (mag. 7.8) remains in *Pisces*.

The path of the Sun and the planets along the ecliptic in November.

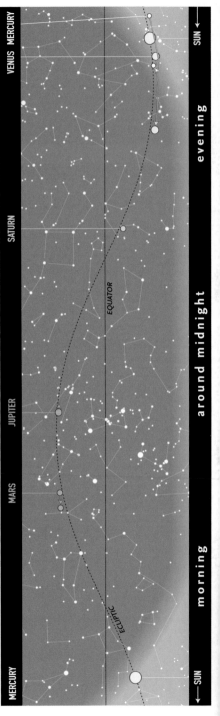

Calendar for November

01	12:47	New Moon
03		Summer Time ends
03	07:36	Mercury (mag. -0.3) 2.1°N of the Moon
04	01:06	Antares 0.1°N of the Moon
05	00:15	Venus (mag. -4.0) 3.1°N of the Moon
06-30		Leonid meteor shower
09	05:55	First Quarter
11	01:43	Saturn (mag. 0.9) 0.1°S of the Moon
12-13		Northern Taurid meteor shower Maximum
12	02:25	Neptune (mag. 7.8) 0.6°S of the Moon
14	11:16	Moon at perigee = 360,109 km
15	21:28	Full Moon
16	01:13	Uranus (mag. 5.6) 4.4°S of the Moon
16	08:09	Mercury at greatest elongation (22.6°E, mag. -0.3)
17	02:02	Aldebaran 10.3°S of the Moon
17	02:45	Uranus at opposition (mag. 5.6)
17	14:53	Jupiter (mag. -2.8) 5.5°S of the Moon
18		Leonid meteor shower maximum
20	02:44	Pollux 1.9°N of the Moon
20	21:09	Mars (mag. -0.3) 2.4°S of the Moon
22	21:29	Regulus 2.7°S of the Moon
23	01:28	Last Quarter
26	11:56	Moon at apogee = 405,314 km
27	12:16	Spica 0.4°S of the Moon

Evening 5:30 p.m.

November 4 • The Moon is close to Venus and surrounded by Antares, Sabik and Kaus Australis (ε Sgr).

Evening 5:30 p.m.

November 10 • The Moon with Saturn, high in the southeast. Fomalhaut is nearby.

Morning 5 a.m.

November 15-16 • The Moon passes Hamal, Uranus (mag. 5.6), the Pleiades and Aldebaran.

Evening 9 p.m.

November 15-17 • The Moon passes Uranus (mag. 5.6), the Pleiades, Aldebaran, Jupiter and Elnath.

Evening 11 p.m.

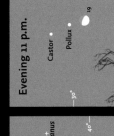

November 19-20 • The Moon passes Castor, Pollux and Mars.

Morning 5 a.m.

November 27 • Near the east-southeast, the Moon and Spica are close together.

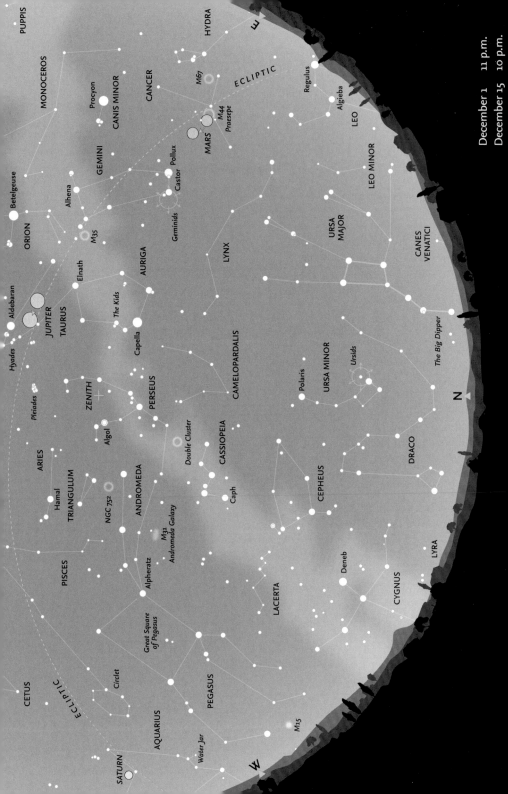

December 1   11 p.m.
December 15   10 p.m.

December – Looking North

Ursa Major has now swung around and is starting to "climb" in the east. The fainter stars in the southern part of the constellation are now fully in view. The other bear, **Ursa Minor**, "hangs" below **Polaris** in the north. Directly above it is the faint constellation of **Camelopardalis**, with the other inconspicuous circumpolar constellation, **Lynx**, to its east. **Vega** (α Lyrae) is now below the horizon in the northwest, but **Deneb** (α Cygni) and most of **Cygnus** remain visible farther west. In the east, **Regulus** (α Leonis) and the constellation of **Leo** are beginning to rise above the horizon. **Cancer** stands high in the east, with **Gemini** even higher in the sky. **Perseus** is at the zenith, with **Auriga** and **Capella** between it and Gemini. Because it is so high in the sky, now is a good time to examine the star clouds of the fainter portion of the Milky Way, between **Cassiopeia** in the west to Gemini and **Orion** in the east.

Meteors

There is one significant meteor shower in December (the last major shower of the year). This is the **Geminid** shower, which is visible over the period December 4–20 and comes to maximum on December 14–15, when the Moon is Full, so conditions are particularly unfavorable. It is one of the most active showers of the year, and in some years is the strongest, with a peak rate of around 100 meteors per hour. It is the one major shower that shows good activity before midnight. The meteors have been found to have a much higher density than other meteors (which are derived from cometary material). It was eventually established that the Geminids and the asteroid Phaethon had similar orbits. So the Geminids are assumed to consist of denser, rocky material. They are slower than most other meteors and often appear to last longer. The brightest often break up into numerous luminous fragments that follow similar paths across the sky. There is a second shower: the **Ursids**, active December 17–26, peaking on December 23,

with a rate at maximum of 5–10, occasionally rising to 25 per hour. Maximum in 2024 occurs when the Moon is around Last Quarter, so conditions are moderately favorable. The parent body is Comet 8P/ Tuttle.

The constellation of Andromeda largely consists of a line of bright stars running northeast from α Andromedae, Alpheratz (bottom right), which is one of the stars forming the Great Square of Pegasus. The small constellation of Triangulum appears on the left-hand (eastern) side, below Andromeda.

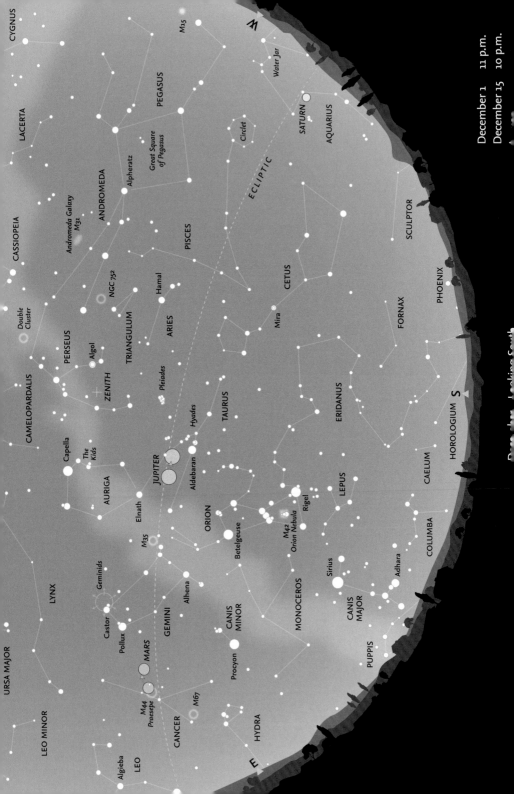

December 1   11 p.m.
December 15 10 p.m.

December, Looking South

December – Looking South

The fine open cluster of the *Pleiades* is due south around 10 p.m., high in the sky, with the *Hyades* cluster, **Aldebaran** and the rest of *Taurus* clearly visible to the east. *Auriga* (with **Capella**) and *Gemini* (with **Castor** and **Pollux**) are both well placed for observation. **Orion** has made a welcome return to the winter sky, and both *Canis Minor* (with **Procyon**) and *Canis Major* (with **Sirius**, the brightest star in the sky) are now well above the horizon. The small, poorly known constellation of *Lepus* lies to the south of Orion, with **Columba** closer to the horizon. The northernmost portion of *Eridanus* is clearly visible in the south. In the west, **Aquarius** has now disappeared, and *Cetus* is becoming lower, but *Pisces* is still easily seen, as are the constellations of *Aries*, *Triangulum* and *Andromeda* above it. The Great Square of *Pegasus* is starting to plunge down toward the western horizon, and because of its orientation on the sky, appears more like a large diamond, standing on one point, than a square.

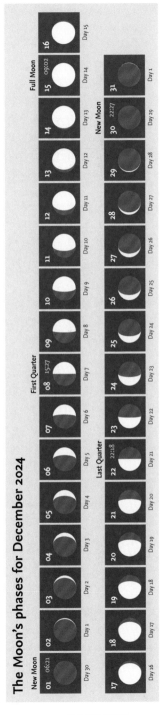

The constellation of Taurus contains two contrasting open clusters: the compact Pleiades, with its striking blue-white stars, and the more scattered, "V"-shaped Hyades, which are much closer to us. Orange Aldebaran (α Tauri) is not related to the Hyades, but lies between it and the Earth.

The Moon's phases for December 2024

New Moon

01 06:21	02	03	04	05	06	07	08 15:27
Day 30	Day 1	Day 2	Day 3	Day 4	Day 5	Day 6	Day 7

First Quarter

09	10	11	12	13	14	15 09:02	16
Day 8	Day 9	Day 10	Day 11	Day 12	Day 13	Day 14	Day 15

Full Moon

17	18	19	20	21	22 22:18	23	24
Day 16	Day 17	Day 18	Day 19	Day 20	Day 21	Day 22	Day 23

Last Quarter

25	26	27	28	29	30 22:27	31	
Day 24	Day 25	Day 26	Day 27	Day 28	Day 29	Day 1	

New Moon

December – Moon and Planets

The Moon

On December 1, the Moon is 0.1°S of *Antares* in *Scorpius*. The next day it is 5.0°S of *Mercury* (too close to the Sun to be visible). By December 4 the Moon is 2.3°S of *Venus* (mag.-4.2) in *Sagittarius*. On December 8 it is 0.3°N of *Saturn* in *Aquarius*. The Moon passes 0.8°N of *Neptune* on December 9 and 4.4°N of *Uranus* on December 13. The next day, December 14, the Moon is 10.3°N of *Aldebaran* and later, 5.5°N of *Jupiter* (mag.-2.8). By December 17 the Moon is 2.1°S of *Pollux*. On December 18 the Moon is 0.9°N of *Mars*, which has begun to retrograde in *Cancer*. On December 20, the Moon passes 2.4°N of *Regulus*. On December 24 it is 0.2°N of *Spica* and on December 28 0.1°S of *Antares*. One day before New, the Moon is 6.4°S of *Mercury*.

The planets

Mercury reaches western elongation on December 25 (see the diagram on page 22). *Venus* (mag. -4.2 to -4.5) moves rapidly across Capricornus. *Mars* begins retrograde motion on December 9. *Jupiter* (mag.-2.8) comes to opposition on December 7. *Saturn* (mag. 1.0 to 1.1) is in *Aquarius*. *Uranus* (mag. 5,6) is in *Taurus*, and *Neptune* (mag. 7.9) remains in *Pisces*, where it has been all year.

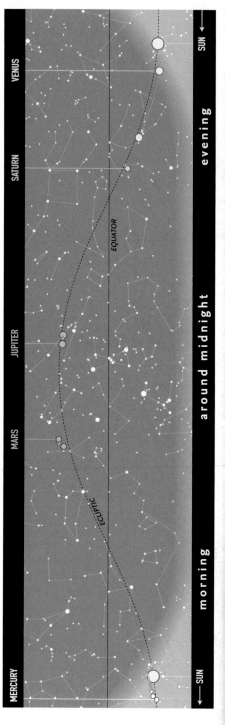

The path of the Sun and the planets along the ecliptic in December.

01	06:21	New Moon
01	07:26	Antares 0.1°N of the Moon
02	02:09	Mercury (mag. 2.7) 5.0°N of the Moon
04–20		Geminid meteor shower
04	22:40	Venus (mag. -4.2) 2.3°N of the Moon
07	20:58	Jupiter at opposition (mag. -2.8)
08	09:19	Saturn (mag. 1.0) 0.3°S of the Moon
08	15:27	First Quarter
09	09:19	Neptune (mag. 7.9) 0.8°S of the Moon
12	13:20	Moon at perigee = 365,361 km
13	09:34	Uranus (mag. 5.6) 4.4°S of the Moon
14–15		Geminid meteor shower maximum
14	05:31	Minor planet (15) Eunomia at opposition (mag. 8.0)
14	12:30	Aldebaran 10.3°S of the Moon
14	19:32	Jupiter (mag. -2.8) 5.5°S of the Moon
15	09:02	Full Moon
17–26		Ursid meteor shower
17	12:49	Pollux 2.1°N of the Moon
18	08:49	Mars (mag. -0.9) 0.9°S of the Moon
20	06:18	Regulus 2.4°S of the Moon
21	09:21	Northern winter solstice
22	22:18	Last Quarter
23		Ursid meteor shower maximum
24	07:25	Moon at apogee = 404,485 km
24	20:11	Spica 0.2°S of the Moon
25	02:30	Mercury at greatest elongation (22.0°W, mag. -0.4)
28	15:17	Antares 0.1°N of the Moon
29	04:21	Mercury (mag. -0.4) 6.4°N of the Moon

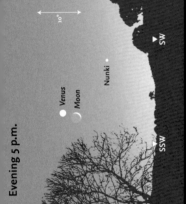

Evening 5 p.m.

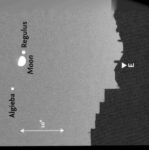

Evening 8 p.m.

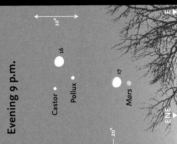

Evening 23:30 p.m.

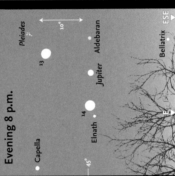

Evening 9 p.m.

Evening 8 p.m.

December 4 • In the evening twilight, the narrow crescent Moon is close to the bright Venus. Nunki (σ Sgr) is closer to the horizon and farther west.

December 7–8 • The Moon passes Saturn, high in the southwest. Fomalhaut is almost twenty degrees lower.

December 13–14 • The Moon passes Jupiter. The Pleiades, Capella, Elnath,

December 16–17 • The Moon passes Castor, Pollux, and Mars.

December 19 • The Moon is between Regulus and Algieba,

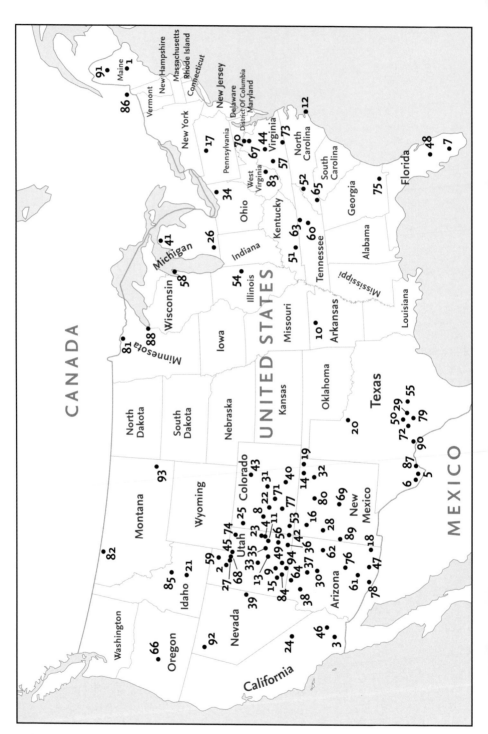

Dark Sky Sites

International Dark Sky Association Sites

The *International Dark-Sky Association* (IDA) recognises various categories of sites that offer areas where the sky is dark at night, free from light pollution and particularly suitable for astronomical observing. There are numerous sites in North America, shown on the map and listed here.

Details of the IDA are at: https://www.darksky.org/. Information on the various categories and individual sites are at: https://www.darksky.org/our-work/conservation/idsp/

Many of these sites have major observatories or other facilities available for public observing (often at specific dates or times).

Parks

1 AMC Maine Woods (ME)
2 Antelope Island State Park (UT)
3 Anza-Borrego Desert State Park (CA)
4 Arches National Park (UT)
5 Big Bend National Park (TX)
6 Big Bend Ranch State Park (TX)
7 Big Cypress National Preserve (FL)
8 Black Canyon of the Gunnison National Park (CO)
9 Bryce Canyon National Park (UT)
10 Buffalo National River (AR)
11 Canyonlands National Park (UT)
12 Cape Lookout National Seashore (NC)
13 Capitol Reef National Park (UT)
14 Capulin Volcano National Monument (NM)
15 Cedar Breaks National Monument (UT)
16 Chaco Culture National Historical Park (NM)
17 Cherry Springs State Park (PA)
18 Chiricahua National Monument (AZ)
19 Clayton Lake State Park (NM)
20 Copper Breaks State Park (TX)
21 Craters Of The Moon National Monument (ID)
22 Curecanti National Recreation Area (CO)
23 Dead Horse Point State Park (UT)
24 Death Valley National Park (CA)
25 Dinosaur National Monument (CO)
26 Dr T.K. Lawless County Park (MI)
27 East Canyon State Park (UT)
28 El Morro National Monument (NM)
29 Enchanted Rock State Natural Area (TX)
30 Flagstaff Area National Monuments (AZ)
31 Florissant Fossil Beds National Monument (CO)
32 Fort Union National Monument (NM)

33 Fremont Indian State Park (UT)
34 Geauga Observatory Park (OH)
35 Goblin Valley State Park (UT)
36 Goosenecks State Park (UT)
37 Grand Canyon National Park (AZ)
38 Grand Canyon-Parashant National Monument (AZ)
39 Great Basin National Park (NV)
40 Great Sand Dunes National Park and Preserve (CO)
41 Headlands (MI)
42 Hovenweep National Monument (UT)
43 Jackson Lake State Park (CO)
44 James River State Park (VA)
45 Jordanelle State Park (UT)
46 Joshua Tree National Park (CA)
47 Kartchner Caverns State Park (AZ)
48 Kissimmee Prairie Preserve State Park (FL)
49 Kodachrome Basin State Park (UT)
50 Lyndon B. Johnson National Historical Park (TX)
51 Mammoth Cave National Park (KY)
52 Mayland Earth to Sky Park & Bare Dark Sky Observatory (NC)
53 Mesa Verde National Park (CO)
54 Middle Fork River Forest Preserve (IL)
55 Milton Reimers Ranch Park (TX)
56 Natural Bridges National Monument (UT)
57 Natural Bridge State Park (VA)
58 Newport State Park (WI)
59 North Fork Park (UT)
60 Obed Wild and Scenic River (TN)
61 Oracle State Park (AZ)
62 Petrified Forest National Park (AZ)
63 Pickett CCC Memorial State Park & Pogue Creek Canyon State Natural Area (TN)
64 Pipe Spring National Monument (AZ)
65 Pisgah Astronomical Research Institute (NC)

66 Prineville Reservoir State Park (OR)
67 Rappahannock County Park (VA)
68 Rockport State Park (UT)
69 Salinas Pueblo Missions National Monument (NM)
70 Sky Meadows State Park (VA)
71 Slumgullion Center (CO)
72 South Llano River State Park (TX)
73 Staunton River State Park (VA)
74 Steinaker State Park (UT)
75 Stephen C. Foster State Park (GA)
76 Tonto National Monument (AZ)
77 Top of the Pines (CO)
78 Tumacácori National Historical Park (AZ)
79 UBarU Camp and Retreat Center (TX)
80 Valles Caldera National Preserve (NM)
81 Voyageurs National Park (MN)
82 Waterton-Glacier International Peace Park (Canada/MT)
83 Watoga State Park (WV)
84 Zion National Park (UT)

Reserves

85 Central Idaho (ID)
86 Mont-Mégantic (Québec)

Sanctuaries

87 Black Gap Wildlife Management Area (TX)
88 Boundary Craters Canoe Area Wilderness (MN)
89 Cosmic Campground (NM)
90 Devils River State Natural Area – Del Norte Unit (TX)
91 Katahdin Woods and Waters National Monument (ME)
92 Massacre Rim (NV)
93 Medicine Rocks State Park (MT)
94 Rainbow Bridge National Monument (UT)

RASC Recognized Dark-Sky Sites

Canadian Dark-Sky Sites

The Royal Astronomical Society of Canada (RASC) has developed formal guidelines and requirements for three types of light-restricted protected areas: Dark-Sky Preserves, Urban Star Parks and Nocturnal Preserves. The focus of the Canadian Program is primarily to protect the nocturnal environment; therefore, the outdoor lighting requirements are the most stringent, but also the most effective. Canadian Parks and other areas that meet these guidelines and successfully apply for one of these designations are officially recognised. Many parks across Canada have been designated in recent years – see the list below and the RASC website: https://www.rasc.ca/dark-sky-site-designations.

Dark-Sky Preserves

1 Torrance Barrens Dark-Sky Preserve (ON)
2 McDonald Park Dark-Sky Park (BC)
3 Cypress Hills Inter-Provincial Park Dark-Sky Preserve (SK/AB)
4 Point Pelee National Park (ON)
5 Beaver Hills and Elk Island National Park (AB)
6 Mont-Mégantic International Dark-Sky Preserve (QC)
7 Manitoulin Eco Park (ON)
8 Grasslands National Park (SK)
9 Bruce Peninsula National Park (ON)
10 Kouchibouguac National Park (NB)
11 Mount Carleton Provincial Park (NB)
12 Kejimkujik National Park (NS)
13 Fundy National Park (NB)
14 Jasper National Park Dark-Sky Preserve (AB)
15 Bluewater Outdoor Education Centre – Wiarton (ON)
16 Wood Buffalo National Park (AB)
17 North Frontenac Township (ON)
18 Lakeland Provincial Park and Provincial Recreation Area (AB)
19 Killarney Provincial Park (ON)
20 Terra Nova National Park (NL)
21 Au Diable Vert (QC)
22 Lake Superior Provincial Park (ON)

Urban Star Parks

23 Irving Nature Park (NB)
24 Cattle Point, Victoria (BC)

Nocturnal Preserves

25 Ann and Sandy Cross Conservation Area (AB)
26 Old Man on His Back Ranch (SK)

Glossary and Tables

aphelion   The point on an orbit that is farthest from the Sun.

apogee   The point on its orbit at which the Moon is farthest from the Earth.

appulse   The apparently close approach of two celestial objects; two planets, or a planet and star.

astronomical unit (AU) The mean distance of the Earth from the Sun, 149,597,870 km.

celestial equator The great circle on the celestial sphere that is in the same plane as the Earth's equator.

celestial sphere The apparent sphere surrounding the Earth on which all celestial bodies (stars, planets, etc.) seem to be located.

conjunction   The point in time when two celestial objects have the same celestial longitude. In the case of the Sun and a planet, superior conjunction occurs when the planet lies on the far side of the Sun (as seen from Earth). For Mercury and Venus, inferior conjunction occurs when they pass between the Sun and the Earth.

direct motion   Motion from west to east on the sky.

ecliptic   The apparent path of the Sun across the sky throughout the year. Also: the plane of the Earth's orbit in space.

elongation   The point at which an inferior planet has the greatest angular distance from the Sun, as seen from Earth.

equinox   The two points during the year when night and day have equal duration. Also: the points on the sky at which the ecliptic intersects the celestial equator. The vernal (spring) equinox is of particular importance in astronomy.

gibbous   The stage in the sequence of phases at which the illumination of a body lies between half and full. In the case of the Moon, the term is applied to phases between First Quarter and Full, and between Full and Last Quarter.

inferior planet   Either of the planets Mercury or Venus, which have orbits inside that of the Earth.

magnitude   The brightness of a star, planet or other celestial body. It is a logarithmic scale, where larger numbers indicate fainter brightness. A difference of 5 in magnitude indicates a difference of 100 in actual brightness, thus a first-magnitude star is 100 times as bright as one of sixth magnitude.

meridian   The great circle passing through the North and South Poles of a body and the observer's position; or the corresponding great circle on the celestial sphere that passes through the North and South Celestial Poles and also through the observer's zenith.

nadir   The point on the celestial sphere directly beneath the observer's feet, opposite the zenith.

occultation   The disappearance of one celestial body behind another, such as when stars or planets are hidden behind the Moon.

opposition   The point on a superior planet's orbit at which it is directly opposite the Sun in the sky.

perigee   The point on its orbit at which the Moon is closest to the Earth.

perihelion   The point on an orbit that is closest to the Sun.

retrograde motion Motion from east to west on the sky.

superior planet   A planet that has an orbit outside that of the Earth.

vernal equinox   The point at which the Sun, in its apparent motion along the ecliptic, crosses the celestial equator from south to north. Also known as the First Point of Aries.

zenith   The point directly above the observer's head.

zodiac   A band, stretching 8° on either side of the ecliptic, within which the Moon and planets appear to move. It consists of 12 equal areas, originally named after the constellation that once lay within it.

The Constellations

There are 88 constellations covering the whole of the celestial sphere, but 16 of these in the southern hemisphere can never be seen (even in part) from a latitude of 40°N, so are omitted from this table. The names themselves are expressed in Latin, and the names of stars are frequently given by Greek letters (see next page) followed by the genitive of the constellation name. The genitives and English names of the various constellations are included.

Name	Genitive	Abbr.	English name
Andromeda	Andromedae	And	Andromeda
Antlia	Antliae	Ant	Air Pump
Aquarius	Aquarii	Aqr	Water Bearer
Aquila	Aquilae	Aql	Eagle
Ara	Arae	Ara	Altar
Aries	Arietis	Ari	Ram
Auriga	Aurigae	Aur	Charioteer
Boötes	Boötis	Boo	Herdsman
Caelum	Caeli	Cae	Burin (Chisel)
Camelopardalis	Camelopardalis	Cam	Giraffe
Cancer	Cancri	Cnc	Crab
Canes Venatici	Canum Venaticorum	CVn	Hunting Dogs
Canis Major	Canis Majoris	CMa	Big Dog
Canis Minor	Canis Minoris	CMi	Little Dog
Capricornus	Capricorni	Cap	Sea Goat
Cassiopeia	Cassiopeiae	Cas	Cassiopeia
Centaurus	Centauri	Cen	Centaur
Cepheus	Cephei	Cep	Cepheus
Cetus	Ceti	Cet	Whale
Columba	Columbae	Col	Dove
Coma Berenices	Comae Berenices	Com	Berenice's Hair
Corona Australis	Coronae Australis	CrA	Southern Crown
Corona Borealis	Coronae Borealis	CrB	Northern Crown
Corvus	Corvi	Crv	Crow
Crater	Crateris	Crt	Cup
Cygnus	Cygni	Cyg	Swan
Delphinus	Delphini	Del	Dolphin
Draco	Draconis	Dra	Dragon
Equuleus	Equulei	Equ	Little Horse
Eridanus	Eridani	Eri	River Eridanus
Fornax	Fornacis	For	Furnace
Gemini	Geminorum	Gem	Twins
Grus	Gruis	Gru	Crane
Hercules	Herculis	Her	Hercules
Horologium	Horologii	Hor	(Pendulum) Clock
Hydra	Hydrae	Hya	Water Snake

Name	Genitive	Abbr.	English name
Indus	Indi	Ind	Indian
Lacerta	Lacertae	Lac	Lizard
Leo	Leonis	Leo	Lion
Leo Minor	Leonis Minoris	LMi	Little Lion
Lepus	Leporis	Lep	Hare
Libra	Librae	Lib	Scales
Lupus	Lupi	Lup	Wolf
Lynx	Lyncis	Lyn	Lynx
Lyra	Lyrae	Lyr	Lyre
Microscopium	Microscopii	Mic	Microscope
Monoceros	Monocerotis	Mon	Unicorn
Norma	Normae	Nor	Level (Square)
Ophiuchus	Ophiuchi	Oph	Serpent Bearer
Orion	Orionis	Ori	Orion
Pegasus	Pegasi	Peg	Pegasus
Perseus	Persei	Per	Perseus
Phoenix	Phoenicis	Phe	Phoenix
Pisces	Piscium	Psc	Fishes
Piscis Austrinus	Piscis Austrini	PsA	Southern Fish
Puppis	Puppis	Pup	Stern
Pyxis	Pyxidis	Pyx	Compass
Sagitta	Sagittae	Sge	Arrow
Sagittarius	Sagittarii	Sgr	Archer
Scorpius	Scorpii	Sco	Scorpion
Sculptor	Sculptoris	Scl	Sculptor
Scutum	Scuti	Sct	Shield
Serpens	Serpentis	Ser	Serpent
Sextans	Sextantis	Sex	Sextant
Taurus	Tauri	Tau	Bull
Telescopium	Telescopii	Tel	Telescope
Triangulum	Trianguli	Tri	Triangle
Ursa Major	Ursae Majoris	UMa	Great Bear
Ursa Minor	Ursae Minoris	UMi	Lesser Bear
Vela	Velorum	Vel	Sails
Virgo	Virginis	Vir	Virgin
Vulpecula	Vulpeculae	Vul	Fox

The Greek Alphabet

α	Alpha	ε	Epsilon	ι	Iota	ν	Nu	ρ	Rho	φ (φ)	Phi
β	Beta	ζ	Zeta	κ	Kappa	ξ	Xi	σ (ς)	Sigma	χ	Chi
γ	Gamma	η	Eta	λ	Lambda	ο	Omicron	τ	Tau	ψ	Psi
δ	Delta	θ (ϑ)	Theta	μ	Mu	π	Pi	υ	Upsilon	ω	Omega

Some common asterisms

Belt of Orion	δ, ε, and ζ Orionis
Big Dipper	α, β, γ, δ, ε, ζ and η Ursae Majoris
Cat's Eyes	λ and υ Scorpii
Circlet	γ, θ, ι, λ and κ Piscium
Guards (or Guardians)	β and γ Ursae Minoris
Head of Cetus	α, γ, ξ², μ and λ Ceti
Head of Draco	β, γ, ξ and ν Draconis
Head of Hydra	δ, ε, ζ, η, ρ and σ Hydrae
Keystone	ε, ζ, η and π Herculis
Kids	ζ and η Aurigae
Little Dipper	β, γ, η, ζ, ε, δ and α Ursae Minoris
Lozenge	= Head of Draco
Milk Dipper	ζ, γ, σ, φ and λ Sagittarii
Plough or Big Dipper	α, β, γ, δ, ε, ζ and η Ursae Majoris
Pointers	α and β Ursae Majoris
Sickle	α, η, γ, ζ, μ and ε Leonis
Square of Pegasus	α, β and γ Pegasi with α Andromedae
Sword of Orion	θ and ι Orionis
Teapot	γ, ε, δ, λ, φ, σ, τ and ζ Sagittarii
Wain (or Charles' Wain)	= Big Dipper
Water Jar	γ, η, κ and ζ Aquarii
Y of Aquarius	= Water Jar

Acknowledgements

Denis Buczynski, Portmahomack, Ross-shire – p.32 (Fireball), p.35 (Quadrantid fireball)
Stephen Edberg – pp.35, 53, 89, 91 (Constellation photographs)
Akira Fuji – p.21 (Lunar eclipse)
Jens Hackmann, Bad Mergentheim, Germany – p.77 (Perseid fireball)
Bernhard Hubl – pp.49, 53, 55, 71, 85, 95 (Constellation photographs)
Nick James – p.29 (Comet NEOWISE)
Damian Peach – p.23 (Mars photo)
peresanz/Shutterstock – p.37 (Orion)
Ken Sperber, California – p.83 (Double Cluster)
Wil Tirion, Capelle aan den IJssel, The Netherlands – p.43, 67, 101 (Constellation photographs)
Alan Tough, Nairn, Highland, Scotland – p.65 (Noctilucent clouds)
For the 2024 Guide, specialist editorial support was provided by Jessica Lee, Astronomy Education Officer, and Patricia Skelton, Senior Astronomy Manager: Education & Outreach at Royal Observatory Greenwich.

Further Information

Books

Bone, Neil (1993), *Observer's Handbook: Meteors*, George Philip, London & Sky Publ. Corp., Cambridge, Mass.

Cook, J., ed. (1999), *The Hatfield Photographic Lunar Atlas*, Springer-Verlag, New York

Dunlop, Storm (2006), *Wild Guide to the Night Sky*, Harper Perennial, New York & Smithsonian Press, Washington D.C.

Dunlop, Storm (2012), *Practical Astronomy*, 2nd edn, Firefly, Buffalo

Dunlop, Storm, Rükl, Antonin & Tirion, Wil (2005), *Collins Atlas of the Night Sky*, HarperCollins, London & Smithsonian Press, Washington D.C.

Grego, Peter (2016), *Moon Observer's Guide*, Firefly, Richmond Hill

Heifetz, Milton D. & Tirion, Wil (2017), *A Walk through the Heavens*, 4th edn, Cambridge University Press, Cambridge

Mellinger, Axel & Hoffmann, Susanne (2005), *The New Atlas of the Stars*, Firefly, Richmond Hill

O'Meara, Stephen J. (2008), *Observing the Night Sky with Binoculars*, Cambridge University Press, Cambridge

Pasachoff, Jay M. (1999), *Peterson Field Guides: Stars and Planets*, 4th edn., Houghton Mifflin, Boston

Ridpath, Ian (2018), *Star Tales*, 2nd edn, Lutterworth Press, Cambridge, UK

Ridpath, Ian, ed. (2003), *Oxford Dictionary of Astronomy*, 2nd edn, Oxford University Press, Oxford

Ridpath, Ian, ed. (2004), *Norton's Star Atlas*, 20th edn, Pi Press, New York

Ridpath, Ian & Tirion, Wil (2004), *Collins Gem – Stars*, HarperCollins, London

Ridpath, Ian & Tirion, Wil (2017), *Collins Pocket Guide Stars and Planets*, 5th edn, HarperCollins, London

Ridpath, Ian & Tirion, Wil (2019), *The Monthly Sky Guide*, 10th edn, Dover Publications, New York

Rükl, Antonín (1990), *Hamlyn Atlas of the Moon*, Hamlyn, London & Astro Media Inc., Milwaukee

Rükl, Antonín (2004), *Atlas of the Moon*, Sky Publishing Corp., Cambridge, Mass.

Scagell, Robin (2015), *Firefly Complete Guide to Stargazing*, Firefly, Richmond Hill

Scagell, Robin & Frydman, David (2014), *Stargazing with Binoculars*, Firefly, Richmond Hill

Scagell, Robin (2014), *Stargazing with a Telescope*, Firefly, Richmond Hill

Stimac, Valerie (2019), *A Practical Guide to Astrotourism*, Lonely Planet

Sky & Telescope (2017), *Astronomy 2018*, Sky Publishing Corp., Cambridge, Mass.

Tirion, Wil (2011), *Cambridge Star Atlas*, 4th edn, Cambridge University Press, Cambridge

Tirion, Wil & Sinnott, Roger (1999), *Sky Atlas 2000.0*, 2nd edn, Sky Publishing Corp., Cambridge, Mass. & Cambridge University Press, Cambridge

Journals

Astronomy, Astro Media Corp., 21027 Crossroads Circle, P.O. Box 1612, Waukesha, WI 53187-1612.
http://astronomynow.com

Sky & Telescope, Sky Publishing Corp., Cambridge, MA 02138-1200.
http://www.skyandtelescope.com/

Societies

American Association of Variable Star Observers (AAVSO), 49 Bay State Rd., Cambridge, MA 02138.
Although primarily concerned with variable stars, the AAVSO also has a solar section.
http://www.aavso.org

American Astronomical Society (AAS), 1667 K Street NW, Suite 800, Washington, DC 20006, New York.
http://aas.org/

American Meteor Society (AMS), Geneseo, New York.
http://www.amsmeteors.org/

Association of Lunar and Planetary Observers (ALPO), ALPO Membership Secretary/Treasurer, P.O. Box 13456, Springfield, IL 62791-3456.
An organization concerned with all forms of amateur astronomical observation, not just the Moon and planets, with numerous coordinated observing sections.
http://alpo-astronomy.org/

Astronomical League (AL), 9201 Ward Parkway Suite #100, Kansas City, MO 64114.
An umbrella organization consisting of over 240 local amateur astronomical societies across the United States.
https://www.astroleague.org/